Post-Kyoto International Climate Policy

The **Harvard Project on International Climate Agreements** is a global, multi-disciplinary effort intended to help identify the key design elements of a scientifically sound, economically rational, and politically pragmatic post-2012 international policy architecture for addressing the threat of climate change. It has commissioned leading scholars to examine a uniquely wide range of core issues that must be addressed if the world is to reach an effective agreement on a successor regime to the Kyoto Protocol. The purpose of the project is not to become an advocate for any single policy but to present the best possible information and analysis on the full range of options concerning mitigation, adaptation, technology, and finance. The main findings of the Harvard Project are reported in this accessible volume for policymakers, prepared by project leaders Joseph E. Aldy and Robert N. Stavins.

A companion volume with a more detailed account of the research is published separately as *Post-Kyoto International Climate Policy: Implementing Architectures for Agreement.*

Joseph E. Aldy is Fellow at Resources for the Future in Washington, DC. He also served on the staff of the President's Council of Economic Advisers, where he was responsible for climate change policy from 1997 to 2000.

Robert N. Stavins is Albert Pratt Professor of Business and Government at the John F. Kennedy School of Government at Harvard University. He is also Director of the Harvard Environmental Economics Program and Chairman of the Kennedy School's Environment and Natural Resources Faculty Group.

Post-Kyoto International Climate Policy

Summary for Policymakers

Joseph E. Aldy and Robert N. Stavins

Research from the Harvard Project on
International Climate Agreements

CAMBRIDGE UNIVERSITY PRESS

Cambridge, New York, Melbourne, Madrid, Cape Town, Singapore, São Paulo, Delhi

Cambridge University Press
The Edinburgh Building, Cambridge CB2 8RU, UK

Published in the United States of America by Cambridge University Press, New York

www.cambridge.org
Information on this title: www.cambridge.org/9780521138000

First published 2009

Printed in the United Kingdom at the University Press, Cambridge

A catalogue record for this publication is available from the British Library

ISBN 978-0-521-13800-0 paperback

To my inspiration, Sarah
J. E. A.

To my loving wife, Joanna
R. N. S.

Contents

International Advisory Board
Harvard Environmental Economics Program

The Harvard Project on International Climate Agreements is an initiative of the Harvard Environmental Economics Program, which develops innovative answers to today's complex environmental issues, through research, teaching, and policy outreach.

Faculty Steering Committee
Harvard Project on International Climate Agreements

Management
Harvard Project on International Climate Agreements

Project Management

Robert Stavins
Director

Robert Stowe
Project Manager

Jason Chapman
Project Coordinator

Tyler Gumpright
Project Assistant

Susan Lynch
Webmaster

Matthew Ranson
Research Assistant

Contributors

Ramgopal Agarwala is a consultant to the World Bank and Asian Development Bank and a Distinguished Fellow at Research and Information System for Developing Countries in New Delhi. He has worked in various senior positions in the World Bank for 25 years, with his last posting to Beijing as the chief economist of the World Bank in China. His most recent research includes articles on international financial architecture and climate change.

Joseph E. Aldy is a Fellow at Resources for the Future. He served on the staff of the President's Council of Economic Advisers from 1997 to 2000, where he was responsible for climate change policy. Dr. Aldy holds a PhD in economics from Harvard University. His research is on international climate change policy architectures; emissions trading programs and other mitigation policies; and the relationship between economic development and greenhouse gas emissions.

Mustafa Babiker holds a BSc in Econometrics and Social Statistics from the University of Kartoum, Sudan, and an MA and PhD in Economics from the University of Colorado-Boulder. He has served as an economist with the MIT Joint Program on the Science and Policy of Global Change and the Arab Planning Institute, and he continues work with the Joint Program on applications of its Emissions Prediction and Policy Analysis model.

Scott Barrett is Lenfest Earth Institute Professor of Natural Resource Economics at Columbia University, in the School of International and Public Affairs. He is the author of *Environment and Statecraft: The Strategy of Environmental Treaty-Making* (2005) and *Why Cooperate? The Incentive to Supply Global Public Goods* (2007). He taught previously at the Johns Hopkins University School of Advanced International Studies and at the London Business School.

Geoffrey J. Blanford currently manages the climate policy research program at the Electric Power Research Institute. His research focuses on energy-economy modeling and the development of integrated assessment tools for application to international climate agreements and technology policy decisions. He has authored several analyses using the MERGE model and holds a PhD in management science and engineering from Stanford University.

Valentina Bosetti holds a PhD in Computational Mathematics and Operations Research from the Università Statale of Milan and an MA in Environmental and Resources Economics from University College London. At the *Fondazione Eni Enrico Mattei,* in Italy, since 2003, she works as a modeler for the Sustainable Development Program, leading the Climate Change Modeling and Policy initiative. She is currently a visiting researcher at the Princeton Environmental Institute.

Katherine Calvin is a Research Economist at the Pacific Northwest National Laboratory's Joint Global Change Research Institute. Dr. Calvin's research focuses on international climate policy regimes, integrated assessment modeling, and the implications of carbon policy on agriculture and land use.

Jing Cao is an assistant professor in the School of Economics and Management, Tsinghua University, Beijing. She is also an affiliated researcher at The Center for China in the World Economy at Tsinghua, at Environmental Development (China Center), and at the Harvard China Project. She has a PhD in Public Policy from Harvard University. Her research focuses on environmental taxation, climate change economics and modeling, productivity measurements, and economic growth.

Carlo Carraro is Professor of Environmental Economics at the University of Venice and Director of Research of the *Fondazione Eni Enrico Mattei.* He is Vice Chair of IPCC Working Group III and Director of the Climate Impacts and Policy Division of the EuroMediterranean Centre on Climate Change. He holds a PhD from Princeton University and is a Research Fellow of CEPR, CESifo and CEPS.

Wenying Chen is Professor in the Institute of Energy, Environment, and Economy, Tsinghua University, Beijing. Her research focuses on energy modeling, integrated assessment models in climate change, carbon capture and storage, and energy systems analysis. Professor Chen has led a number of national and international research projects in the field of energy and climate change.

Leon Clarke is Senior Research Economist at the Pacific Northwest National Laboratory's Joint Global Change Research Institute. Dr. Clarke's research focuses on technology planning for climate change, climate mitigation scenarios, international climate policy, and integrated assessment of climate change.

Richard N. Cooper is Maurits C. Boas Professor of International Economics at Harvard University. He is Vice-Chairman of the Global Development Network and a member of the Trilateral Commission, the Council on Foreign Relations, the Executive Panel of the US Chief of Naval Operations, and the Brookings Panel on Economic Activity. He has served on several occasions in the US Government, most recently as chairman of the National Intelligence Council (1995–1997).

Jae Edmonds is Chief Scientist and Laboratory Fellow at the Pacific Northwest National Laboratory's Joint Global Change Research Institute, Adjunct Professor of Public Policy at the University of Maryland-College Park, and has actively participated in the IPCC. His research in the areas of long-term, global energy, technology, economy, and climate change spans three decades, producing several books and numerous scientific papers and presentations.

A. Denny Ellerman is Senior Lecturer at MIT's Sloan School of Management and an internationally recognized expert on energy and environmental economics with a particular focus on emissions trading. He is a co-author of the leading book on the US SO_2 Trading Program, *Markets for Clean Air*, and co-editor of *Allocation in the European Emissions Trading Scheme*. He earned a PhD in political economy and government from Harvard University.

Carolyn Fischer is Senior Fellow at Resources for the Future in Washington, DC. Her research addresses a variety of environmental policy issues, including climate change mitigation, technological change, international trade and environmental policies, and resource economics. She holds a PhD in Economics from the University of Michigan and a BA in International Relations from the University of Pennsylvania, and she previously served at the White House Council of Economic Advisors.

Jeffrey Frankel is Harpel Professor at Harvard's Kennedy School. He directs the program in International Finance and Macroeconomics at the National Bureau of Economic Research, where he is also on the Business Cycle Dating Committee. He served on President Clinton's Council of Economic Advisers (1996–1999), with responsibility for environmental, international, and macroeconomics. Earlier he was professor of economics at the University of California, Berkeley. His Economics PhD is from MIT.

Daniel S. Hall is Research Associate at Resources for the Future, where his work focuses on climate change policy, including mechanisms for cost containment, the design of offset programs, and legislative analysis. Hall holds a Master of Environmental Science and Management from the Donald Bren School at the University of California, Santa Barbara.

Bård Harstad is Associate Professor at the Kellogg School of Management, Northwestern University. In recent years he has developed theories for international political economy, with a particular focus on international bargaining and the design elements of environmental agreements. His research has been published in *American Economic Review, Quarterly Journal of Economics*, and *Scandinavian Journal of Economics*.

Jiankun He is the Director of the Low Carbon Energy Laboratory at Tsinghua University, China. Professor He's research interests include energy systems engineering and energy modeling, strategic responses to climate change, resource management, and sustainable development. He has been the principal investigator of a number of national key research projects and international collaborative research projects.

Henry D. Jacoby is Professor of Management in the MIT Sloan School of Management and Co-Director of the MIT Joint Program on the Science and Policy of Global Change. He holds a PhD in Economics from Harvard, where he served in the Department of Economics and the Kennedy School of Government. He serves on the Scientific Committee of the International Geosphere-Biosphere Program and on the Climate Research Committee of the National Research Council.

Judson Jaffe is a Vice President at Analysis Group, Inc. He previously spent two years on the staff of the Council of Economic Advisers at the White House, where he provided economic analysis of environmental and energy policy. He received an MPhil in Economics from Cambridge University, and an AB *summa cum laude* in Environmental Science, Public Policy, and Economics from Harvard University.

Larry Karp is Professor of Agricultural and Resource Economics and the Department Chair at the University of California, Berkeley. His research and teaching interests include environmental economics, trade policy, dynamic methods, and industrial organization. He is Associate Editor of the *Journal of Economic Dynamics and Control* and has served as Co-editor of the *Journal of Environmental Economics and Management*. He is a Fellow of the Agricultural and Applied Economics Association.

Andrew Keeler teaches at the John Glenn School of Public Affairs at the Ohio State University and writes on state, national, and international climate change policy. He served as the Senior Staff Economist for Environment at the President's Council of Economic Advisers (2000–2001) where he was a member of the US negotiating team for climate change and a diplomatic representative to OECD meetings on coordinating national sustainability policies.

Robert O. Keohane is Professor of International Affairs, Princeton University. He is the author of *After Hegemony* (1984) and *Power and Governance in a Partially Globalized World* (2002). He is co-author (with Joseph S. Nye, Jr.) of *Power and Interdependence* (third edition 2001) and (with Gary King and Sidney Verba) of *Designing Social Inquiry* (1994). He is a member of the American Academy of Arts and Sciences, the American Philosophical Society, and the National Academy of Sciences.

Page Kyle is a research analyst with the Pacific Northwest National Laboratory's Joint Global Change Research Institute. His research focuses on modeling of greenhouse gas emissions from end-use energy consumption and secondary fuel production, with particular attention to technological development and climate change mitigation.

Michael A. Levi is the David M. Rubenstein Senior Fellow for Energy and the Environment at the Council on Foreign Relations (CFR) and Director of its Program on Energy Security and Climate Change. He was project director for a recent CFR-sponsored independent task force on climate change, and is the author of *On Nuclear Terrorism* (Harvard University Press, 2007) and *The Future of Arms Control* (Brookings Institution Press, 2005).

Warwick J. McKibbin is Professor and Director of the Centre for Applied Macroeconomic Analysis in the College of Business and Economics at the Australian National University. He also holds positions at the Lowy Institute for International Policy and the Brookings Institution. He is a member of the policy Board of the Reserve Bank of Australia. Professor McKibbin received a PhD in Economics from Harvard University in 1986.

Richard D. Morgenstern is Senior Fellow at Resources for the Future and has written widely on climate change mitigation policy. His involvement in the issue reaches back two decades and includes his work at the US EPA, where he directed the Agency's climate change activities and, subsequently, as a member of the State Department's negotiating team for the Kyoto Protocol.

Adele Morris is a fellow and Deputy Director for Climate and Energy Economics at the Brookings Institution. Her economic and natural resource policy experience includes work at the Joint Economic Committee of Congress, the US Treasury, the President's Council of Economic Advisers, and OMB. She was a lead climate negotiator with the US State Department in 2000. She holds a PhD in Economics from Princeton University.

Richard G. Newell is Gendell Professor of Energy and Environmental Economics, Nicholas School of the Environment, Duke University; a Research Associate, National Bureau of Economic Research; and a University Fellow, Resources for the Future. He has served as Senior Economist for energy and environment on the President's Council of Economic Advisers and on several National Academy of Sciences committees related to energy, environment, and climate. His PhD is from Harvard University.

Sergey V. Paltsev is Principal Research Scientist at the Joint Program on the Science and Policy of Global Change at the Massachusetts Institute of Technology, where he has been working since 2002 as the lead modeler in charge of the MIT Emissions Prediction and Policy Analysis (EPPA) model, a multi-regional computable general equilibrium model of the world economy that has been widely used to study climate change policy.

William A. Pizer is the Deputy Assistant Secretary for Environment and Energy at the US Department of the Treasury. Prior to coming to Treasury, and throughout his involvement with the Harvard Project, Pizer was a Senior Fellow at Resources for the Future where his research looked at how the design of environmental policy affects costs and environmental effectiveness, often related to global climate change. He holds a PhD in Economics from Harvard University.

Andrew J. Plantinga is Professor of Agricultural and Resource Economics at Oregon State University. He received a PhD in Agricultural and Resource Economics from the University of California, Berkeley and an MS in Forestry from the University of Wisconsin-Madison. His research on the economics of land use is supported by the National Science Foundation, the US Forest Service, and the US Department of Energy.

Eric A. Posner is Kirkland & Ellis Professor of Law, University of Chicago. He is author of *The Perils of Global Legalism* (University of Chicago, forthcoming); *Terror in the Balance: Security, Liberty and the Courts* (with Adrian Vermeule) (Oxford, 2007); *New Foundations of Cost-Benefit Analysis* (with Matthew Adler) (Harvard, 2006); *The Limits of International Law* (with Jack Goldsmith) (Oxford, 2005); and *Law and Social Norms* (Harvard, 2000).

Kal Raustiala is a professor at UCLA Law School and the UCLA International Institute, where he is also Director of the Ronald W. Burkle Center for International Relations. His previous publications include *The Implementation and Effectiveness of International Environmental Commitments* (MIT, 1998), co-edited with David G. Victor and Eugene Skolnikoff.

John M. Reilly is the Associate Director for Research in the Joint Program on the Science and Policy of Global Change and a Senior Lecturer in the Sloan School at MIT. Prior appointments were with the USDA's Economic Research Service and the US DOE National Laboratories. He holds a PhD in economics from the University of Pennsylvania. His research has focused on the economics of energy, agriculture, and climate change.

Kenneth R. Richards is an associate professor at the School of Public and Environmental Affairs and an adjunct professor at the Maurer School of Law, Indiana University. He holds a PhD in Public Policy and a JD from the University of Pennsylvania. He is associate director of the Richard G. Lugar Center for Renewable Energy in Indianapolis and the Center for Research in Energy and the Environment in Bloomington, Indiana.

Richard G. Richels is Senior Technical Executive for global climate change research at the Electric Power Research Institute and is Adjunct Professor at the Johns Hopkins School of Advanced International Studies. He has served on a number of national and international advisory panels, including committees of the Department of Energy, the Environmental Protection Agency, the National Research Council, and the Intergovernmental Panel on Climate Change.

Thomas F. Rutherford has been Professor of Energy Economics at ETH Zürich since January, 2008. He earned a PhD in Operations Research from Stanford University under the supervision of Alan S. Manne. He subsequently had academic appointments in economics at the University of Western Ontario and the University of Colorado. Professor Rutherford's main research areas concern the formulation and analysis of computational economic equilibrium models.

Akihiro Sawa is Senior Executive Fellow at the 21st Century Public Policy Research Institute, Keidanren, Tokyo, Japan. He was previously Director of Environmental Policy (2001–2003) and Director of Resources and Fuel Policy (2003–2004) for the Ministry of Economy, Trade and Industry of the Government of Japan and a Professor at the Research Center for Advanced Science and Technology, University of Tokyo (2004–2008).

Richard Schmalensee is the Howard W. Johnson Professor of Economics and Management at MIT and Director of the MIT Center for Energy and Environmental Policy Research. He has served as the John C. Head III Dean of the MIT Sloan School of Management (1998–2007) and as the Member of the President's Council of Economic Advisers with responsibility for environmental policy (1989–1991).

Alessandra Sgobbi holds a PhD in Analysis and Governance of Sustainable Development at the School for Advanced Studies in the Venice Foundation. She collaborates with the *Fondazione Eni Enrico Mattei,* in Italy, on various projects in the field of natural resources management and climate change. Currently, she works at the European Commission, EuropeAid Cooperation Office, focusing on development interventions in the fields of sustainable consumption and production, energy efficiency, and the "grey" environment.

E. Somanathan received a PhD in economics from Harvard in 1995 and taught at Emory University and the University of Michigan at Ann Arbor before joining the Indian Statistical Institute, Delhi, where he is Professor in the Planning Unit. His main research interest is in development economics, particularly environmental problems and political economy. He is writing a book on environmental issues in India.

Robert N. Stavins is Albert Pratt Professor of Business and Government, Harvard Kennedy School; Director, Harvard Environmental Economics Program; Director, Harvard Project on International Climate Agreements; University Fellow, Resources for the Future; Research Associate, National Bureau of Economic Research; and Editor, *Review of Environmental Economics and Policy.* He was Chairman, US EPA Environmental Economics Advisory Committee, and Lead Author, Intergovernmental Panel on Climate Change. He holds a PhD in economics from Harvard.

Cass R. Sunstein is the Felix Frankfurter Professor of Law at Harvard Law School. A former attorney-adviser in the Office of Legal Counsel in the Department of Justice, he is author or co-author of more than fifteen books and hundreds of scholarly articles. Sunstein joined the law faculty of the University of Chicago in 1981 and later became the Karl N. Llewellyn Distinguished Service Professor of Jurisprudence at the University.

Massimo Tavoni is a Senior Researcher at the *Fondazione Eni Enrico Mattei,* in Italy, and is now a post-doctoral research fellow at the Princeton Environmental Institute. His research interests include international climate mitigation policies, technological evolution and uncertainty, and the role of tropical deforestation. He holds an MSc in Mathematical Economics from the London School of Economics and a PhD in Political Economics from the Catholic University of Milan.

Fei Teng is Associate Professor at the Institute of Energy, Environment, and Economy, Tsinghua University, Beijing. His research interests include climate policy analysis, energy policy analysis, and technology transfer mechanisms in climate regimes. He is a review expert for the CDM DNA in China and also a member of the Chinese delegation to the UNFCCC and its Kyoto Protocol.

Alexander Thompson is Associate Professor of Political Science at the Ohio State University. He has research interests in the areas of international organizations and US foreign policy. He is the author of *Channels of Power: The UN Security Council and U.S. Statecraft in Iraq* (Cornell University Press, 2009) and articles in various journals, including *International Organization*, the *Journal of Conflict Resolution*, and the *Journal of Legal Studies*.

Takahiro Ueno is a researcher at the Socio-economic Research Center of the Central Research Institute of Electric Power Industry, Japan. He was a Visiting Scholar at Resources for the Future in 2006 and 2007. He has researched international negotiations on climate change, energy and environmental technology policy, international cooperation on energy efficiency, and technology transfer to developing countries.

David G. Victor is Professor at Stanford Law School and Director of the Program on Energy and Sustainable Development; he also serves as Senior Fellow at the Council on Foreign Relations. His current research focuses on the performance of state-controlled oil companies, on global climate protection, and on the emerging global market for coal. His PhD is from the Massachusetts Institute of Technology and his BA is from Harvard University.

Peter J. Wilcoxen is Associate Professor of Economics and Public Administration at the Maxwell School of Syracuse University and a Nonresident Senior Fellow at the Brookings Institution. He has published extensively on energy and environmental policy and is currently a member of the US EPA's Environmental Economics Advisory Committee. He holds a BA in physics from the University of Colorado and a PhD in economics from Harvard University.

Timothy E. Wirth has been President of the United Nations Foundation since its founding in 1998. He represented Colorado in the US House of Representatives from 1975 to 1987 and the US Senate from 1987 to 1993. From 1993 to 1997, he served as the first US Under Secretary of State for Global Affairs, leading the US team preparing for the Kyoto climate negotiations. He received a PhD from Stanford University.

Marshall Wise is Senior Research Scientist at Battelle's Joint Global Change Research Institute at the University of Maryland. Wise is a long-time member of the MiniCAM integrated assessment model development team with expertise in economic modeling and analysis of energy systems, with experience in both broad-scale energy policy analysis and in detailed analysis of the electric power generation sector.

Jinhua Zhao is an associate professor in the Department of Economics and the Department of Agricultural, Food and Resource Economics at Michigan State University. His research interests include applied microeconomic theory, environmental and resource economics, energy economics, and dynamic decision making under uncertainty, among others. He was a co-editor of the *Journal of Environmental Economics and Management* (JEEM) and is on the editorial council of JEEM and the *Review of Development Economics.*

Foreword

Timothy E. Wirth

Washington, DC
February 5, 2009

When Charles Keeling began measuring carbon dioxide at Mauna Loa in 1958, the atmospheric concentration was 315 parts per million (ppm). That number represented an increase of 12.5 percent from the pre-industrial level of 280 ppm. Fifty years later, it has reached 385 ppm, and the rate of increase has doubled.

As the Swedish chemist Svante Arrhenius predicted in 1896, those increased levels of carbon dioxide or CO_2 are warming the surface temperature of the Earth. The results are evident all around us. The world's tropical belt has expanded toward the poles by two degrees of latitude – as much as had been predicted for the entire twenty-first century. The Greenland ice sheet, which holds enough water to raise global sea levels by 20 feet, is melting at an accelerated rate. The Arctic Ocean – engine of the Northern Hemisphere's weather – could be ice-free during the summer within five years.

Civilization was built around the climate we have – along coastlines that may be washed away by storms and rising sea levels; around farmland and forests that will become less productive as water supplies diminish; at elevations cool enough to escape insect-borne disease. Changing the climate puts the very organization of modern societies at risk.

We cannot avoid climate change altogether. The effects of our actions are already clear. For all practical purposes, they are irreversible. We can, however, limit the damage, and toward that end, the world must act – urgently, dramatically, and decisively.

This summary of an important new volume – the product of the Harvard Project on International Climate Agreements – recognizes the gravity and complexity of the climate challenge. It attempts to show the way forward, building on a rich variety of contributions from more than two dozen experts.

Joseph Aldy and Robert Stavins have underscored design elements for a new international climate regime that meet three well-chosen criteria: They must be scientifically sound, economically rational, and politically pragmatic.

Publication could not be more timely. The world is poised at a hinge of history. Civilization's future rests with decisions yet unmade. Hope and fear collide.

Scientists agree that time is running out for concerted action to avert the worst consequences of climate change. The process that was initiated in Rio de Janeiro in 1992, when agreement was reached on the United Nations Framework Convention on Climate Change, must now achieve a new level of commitment. For the essential objective of the Rio treaty – ratified by the United States and nearly every country of the world – was to prevent "dangerous anthropogenic interference with the climate system." Now, physical evidence of climate change suggests that point has already been passed. Some climate scientists say the world must limit atmospheric CO_2 not to 550 ppm (a doubling of pre-industrial levels), or to 450 ppm (the number often associated with a global warming of 2° C), but to 350 ppm – the level passed 20 years ago – to avoid irreversible melting of the Greenland ice sheet and disastrous sea-level rise.

In December 2007, representatives of 187 countries agreed in Bali on a road map to replace the Kyoto Protocol when it runs out in 2012 and more effectively confront climate change over the long term. Ban Ki-moon, the Secretary-General of the United Nations and a new voice of global leadership, has made climate change one of his top priorities at the UN. "Today we are at a crossroads," he said at Bali, "one path towards a comprehensive new climate agreement, and the other towards a betrayal of our planet and our children. The choice is clear."

Ban left the talks, but when they threatened to founder, he returned to urge the negotiators on. They listened, and adopted a two-year plan for reaching a new agreement. With the inauguration of Barack Obama as US President in January 2009, the world's largest economy is prepared to participate constructively again. Many countries are hoping that the United States will be the cavalry riding to the rescue; it remains to be seen whether that hope is too audacious.

What are the key elements of an agreement? The Bali road map identifies four: mitigation, adaptation, technology, and finance. In the parlance of climate negotiations, "mitigation" means reducing greenhouse gas emissions, and "adaptation" means preparing for climate impacts that cannot be avoided. "Technology" refers to the need, not just to develop cleaner ways of producing and using energy, but also to deploy those technologies on an appropriate scale in rich and poor countries alike. "Finance" encompasses both the mechanisms and investment flows that will enable poor countries to adapt and acquire clean energy technologies.

The UN Framework Convention of 1992 established the principle that countries should engage the climate challenge "on the basis of equity and in accordance with their common but differentiated responsibilities and respective capabilities." Developed countries, especially the United States, were expected to lead because over many years they have contributed the most to the buildup of greenhouse gases in the atmosphere. Meaningful engagement of developing countries, especially of rapidly industrializing economies like China and India, is needed also. All countries must be part of the solution, not just the industrialized countries that caused the problem, but the poorest countries that will feel its effects most acutely.

The question of who has what responsibility, and when obligations will kick in, is the central issue in international climate negotiations, and one that will also be critical to the future ratification of any new climate protocol in the United States and around the world. We must be flexible enough to recognize and accept the value of diverse approaches to the climate challenge.

This *Summary for Policy Makers* – and the edited volume it distills – reflects that imperative, drawing on scholars from China, India, Japan, and Australia, as well as Europe and the United States. There are many good ideas here – too many to summarize briefly. Aldy and Stavins, in their valuable synthesis, point to four potential architectures for agreement. In many ways the four can coexist and support each other:

- Binding emissions caps are needed to bring about reductions from major greenhouse gas sources, although some rapidly industrializing countries may have to step up to that responsibility gradually. Using formulas to allocate reductions is a promising approach for avoiding decisions based simply on politics and power.
- Harmonized domestic policies would facilitate effective implementation of emissions cuts and reduce both the cost of compliance and the political resistance to carbon limits.
- A system of harmonized carbon taxes would generate revenues equitably to support a comprehensive climate response.
- Linked national cap-and-trade systems, based on permit auctions implemented "upstream," would do the same.

The relationship between these approaches can be seen by considering how best to encourage technology deployment and economic development.

Solving the climate crisis will require nothing less than a fundamental transformation of global energy systems. In the United States, transportation and electricity generation are the two largest sources of emissions. In rapidly industrializing nations like China and India, power generation, manufacturing, and transportation are the fastest-growing sources. A new generation of climate-friendly technologies will be needed to reduce emissions quickly and at low cost.

The global recession that began in 2008 as a result of turbulence in world financial markets creates new barriers, as well as new opportunities for major new investments in clean energy technologies. Falling commodity prices, especially for oil, have reduced political pressure for immediate action on energy policy even as capital for new projects has become much more difficult to obtain. The need for substantial government spending to revive the economy, on the other hand, provides a potential stimulus to jump-start the transition to new energy technologies.

In the US presidential election of 2008, both major party candidates made investment in renewable energy a centerpiece of their campaigns, reflecting the breadth of bipartisan support for a change in direction. Research and development are not enough, though – new market signals are essential for this technological revolution to succeed. The most important step is to put a price on carbon, either through a cap-and-trade system or a carbon tax. The purpose is not to penalize consumers with higher energy costs. Rather it is to set the rules of the game so that clean technologies can compete with dirty ones – and indeed, out-compete them over time. This will lead to a great wave of innovation, investment, economic development, and job creation.

Serious action by the United States to significantly reduce its emissions is not only the right thing to do; it is also a precondition for US credibility and global leadership on climate. Without it, other countries will have a convenient excuse for inaction.

Key steps to reduce emissions will include increased efficiency, the transformation of the transportation sector through advanced biofuels and plug-in hybrids, and the phase-out of conventional coal-fired power generation. Such steps could become the basis for harmonized national policies – setting, for example, targets for improvement in energy efficiency and deployment of renewable energy – that could be endorsed globally as confidence-building steps toward a new climate agreement.

The US-China relationship is critical to such progress. These countries are the world's two largest emitters of greenhouse gases, and neither accepted any restrictions under the Kyoto protocol. China continues to resist the idea of binding targets, but on its own has set a target of improving the energy efficiency of its economy by an extraordinary 4 percent per year. China has also imposed vehicle fuel economy standards stricter than those of the United States and plans to double its renewable energy capacity (to 15 percent of its overall energy supply mix) by the year 2020. These steps could be a model for other countries and the basis for voluntary targets, globally agreed.

Developing countries, especially China and India, will account for the lion's share of global emissions growth over the coming years. In China alone, as many as 500 million people will join the middle class, gaining access to electricity and motorized transportation, in the next 20 years. In recent years, China has been expanding its coal base at the rate of one large new coal-fired power plant, on average, every week, and India aspires to similar economic growth. Getting these countries to grow cleanly, therefore, is absolutely essential to climate stabilization. The idea of giving handouts to increasingly formidable competitors overseas is politically toxic in many developed countries, but more robust cooperation in areas of mutual interest – such as advancing carbon capture technology for coal plants – would accelerate technology development and deployment to the benefit of all.

Development and clean energy should go hand in hand – the limitations of the dirty energy path are more manifest by the day – but the phrase "technology transfer" has an unfortunate ring. It suggests hand-me-downs from rich countries to poor. Instead, nations that are technology leaders should collaborate on a new international initiative to facilitate cooperation with developing countries on low-cost clean-energy technologies. Working together through regional innovation centers, researchers would adapt these technologies to their countries and help them "leapfrog" over climate-damaging business-as-usual patterns of development, much as the advent of cell phones averted a massive buildout of telecommunications infrastructure.

Managing the climate crisis requires new forms of international cooperation to reduce global emissions and help vulnerable societies adapt. The UN is the appropriate venue for global negotiation, and in many cases the right institution to coordinate and deliver international response measures. The United States can lead this global effort by reducing its own emissions, encouraging other nations to implement bold mitigation policies, spurring technological innovation at home and abroad, speeding adoption of clean energy technologies by rapidly developing nations, and providing adaptation assistance to poor nations.

International climate negotiations are complex – to be successful, they will require political resolve, creative negotiating, innovative policy mechanisms, stronger global institutions, and additional financial resources. None of this will be easy, but a flexible and positive approach can yield results if it focuses – as the Harvard Project does – on solutions that are scientifically sound, economically rational, and politically pragmatic. The world can afford no less. If this volume moves negotiators closer to that goal, the Harvard Project on International Climate Agreements will have provided value indeed.

Joseph E. Aldy and Robert N. Stavins

Diverse aspects of human activity around the world result in greenhouse gas (GHG) emissions that contribute to global climate change. Emissions come from coal-fired power plants in the United States, diesel buses in Europe, rice paddies in Asia, and the burning of tropical forests in South America. These emissions will affect the global climate for generations, because most greenhouse gases reside in the atmosphere for decades to centuries. Thus, the impacts of global climate change pose serious, long-term risks.

Global climate change is the ultimate global-commons problem: Because GHGs mix uniformly in the upper atmosphere, damages are completely independent of the location of emissions sources. Thus, a multinational response is required. To address effectively the risks of climate change, efforts that engage most if not all countries will need to be undertaken. The greatest challenge lies in designing an *international policy architecture* that can guide such efforts. We take "international policy architecture" to refer to the basic nature and structure of an international agreement or other multilateral (or bilateral) climate regime.[2]

[1] We are indebted to the twenty-six research teams of the Harvard Project on International Climate Agreements who have contributed to the Project, this *Summary for Policymakers*, and the complete book (Aldy and Stavins 2009). We are also grateful to the Project's management: Robert Stowe, project manager; Sasha Talcott, communications director; Jason Chapman, project coordinator; Tyler Gumpright, project assistant; Susan Lynch, webmaster; and Matthew Ranson, research assistant. We are particularly grateful to Rob Stowe, who has managed the production of this *Summary for Policymakers* – and the overall Harvard Project – with inspired leadership and unfailing grace and kindness. Marika Tatsutani edited the manuscript with skill and insight. We also express our sincere gratitude to the Doris Duke Charitable Foundation for providing major funding for the Project, and Andrew Bowman for his collaboration, beginning with the Project's conception. We greatly appreciate additional financial support from Christopher Kaneb, the James and Cathleen Stone Foundation, Paul Josefowitz and Nicholas Josefowitz, the Enel Endowment for Environmental Economics at Harvard University, the Belfer Center for Science and International Affairs at the Harvard Kennedy School, and the Mossavar-Rahmani Center for Business and Government at the Harvard Kennedy School.

[2] The need for scholars to focus on the development of a long-term climate policy architecture was first highlighted by Richard Schmalensee: "When time is measured in centuries, the creation of durable institutions and frameworks seems both logically prior to and more important than choice of a particular policy program that will almost surely be viewed as too strong or too weak within a decade" (1998, p. 141).

The Kyoto Protocol to the United Nations Framework Convention on Climate Change (UNFCCC) marked the first meaningful attempt by the community of nations to curb GHG emissions. This agreement, though a significant first step, is not sufficient for the longer-term task ahead. Some observers support the policy approach embodied in Kyoto and would like to see it extended – perhaps with modifications – beyond the first commitment period, which ends in 2012. Others maintain that a fundamentally new approach is required.

Whether one thinks the Kyoto Protocol was a good first step or a bad first step, everyone agrees that a second step is required. A way forward is needed for the post-2012 period. The Harvard Project on International Climate Agreements was launched with this imperative in mind. The Project is a global, multiyear, multi-disciplinary effort intended to help identify the key design elements of a scientifically sound, economically rational, and politically pragmatic post-2012 international policy architecture for addressing the threat of climate change. This *Summary for Policymakers* is a product of the Project's research, the results of which are described in much greater detail in our book, *Post-Kyoto International Climate Policy: Implementing Architectures for Agreement* (Aldy and Stavins 2009).

By "scientifically sound" we mean an international agreement that is consistent with achieving the objective of stabilizing atmospheric concentrations of GHGs at levels that avoid dangerous anthropogenic interference with the global climate. By "economically rational" we mean pursuing an approach or set of approaches that are likely to achieve global targets at minimum cost – that is, cost-effectively. And by "politically pragmatic" we mean a post-Kyoto regime that is likely to bring on board the United States and engage key, rapidly-growing developing countries in increasingly meaningful ways over time. As Tim Wirth emphasizes in his Foreword, these three criteria are essential for identifying a promising and meaningful path forward.

The Project draws upon leading thinkers from academia, private industry, government, and non-governmental organizations (NGOs) around the world. It includes research teams operating in Europe, the United States, China, India, Japan, and Australia, and has benefited from meetings with leaders from business, NGOs, and governments in many more countries.

The Project originated from a May 2006 workshop at which the Harvard Environmental Economics Program brought together twenty-seven leading thinkers from around the world with expertise in economics, law, political science, business, international relations, and the natural sciences. This group developed and refined six policy frameworks, each of which could form the backbone of a new international climate agreement. These six frameworks, which range from a stronger version of the Kyoto Protocol to entirely new approaches, are the subject of our earlier book, published in September 2007 by Cambridge University Press and titled *Architectures for Agreement: Addressing Global Climate Change in the Post-Kyoto World* (Aldy and Stavins 2007). With these proposals as the starting point, the Harvard Project on International Climate Agreements aims to help forge a broad-based consensus on a potential successor to Kyoto.

The first stage of our work, which focused on establishing the importance of considering alternative architectures for the post-2012 period, featured wide-ranging and inclusive discussions of the six proposed alternatives, as well as others not addressed in *Architectures for Agreement*. It also featured meetings with government officials, business leaders, NGOs, and academics around the world. In the second stage of the Project, we focused on developing a small menu of promising frameworks and key design principles, based upon analysis by leading academics from a variety of disciplines – including economics, political science, law, and international relations – as well as ongoing commentary from leading practitioners in the NGO community, private industry, and government. Economic analysis has been supplemented with political analysis of the implications of alternative approaches, as well as legal examinations of the feasibility of various proposals.

From the beginning, there have been no constraints on what might emerge from the Project. We have maintained from the outset that anything is possible – from highly centralized Kyoto-like architectures for all countries to proposals that are outside the context of the UNFCCC, such as proposals for G8+5 or L20 agreements.[3] This *Summary for Policymakers* draws upon the findings of our diverse research initiatives in Australia, China, Europe, India, Japan, and the United States.

Learning from experience: the Kyoto Protocol

It is helpful to reflect on the lessons that can be learned from examining the Kyoto Protocol's strengths and weaknesses. Among the Protocol's strengths

---

[3] The G8 refers to Canada, France, Germany, Italy, Japan, Russia, the United Kingdom, and the United States; in addition, the EU is represented within the G8, but cannot host or chair. The G8+5 refers to the G8 countries plus the 5 leading developing countries – Brazil, China, India, Mexico, and South Africa. The L20 refers to the G8+5 nations plus Australia, Argentina, the European Union, Indonesia, Korea, Saudi Arabia, and Turkey.

is its inclusion of several provisions for market-based approaches that hold promise for improving the cost-effectiveness of a global climate regime. We refer, for example, to the well-known flexibility mechanisms, such as Article 17, which provides for emissions trading among the Annex I countries[4] that take on commitments under the Protocol. More specifically, this provision allows the governments of Annex I countries to trade some of the assigned emission allowances that constitute their country-level targets. Second, the Protocol's Joint Implementation provisions allow for project-level trades among the Annex I countries. Finally, the Protocol established the Clean Development Mechanism (CDM), which provides for the use of project-level emission offsets created in non-Annex I countries (the developing countries of the world) to help meet the compliance obligations of Annex I countries.

A second advantage of the Kyoto Protocol is that it provides flexibility for nations to meet their national emission targets – their commitments – in any way they want. In other words, Article 2 of the Protocol recognizes domestic sovereignty by providing for flexibility at the national level. The political importance of this provision in terms of making it possible for a large number of nations to reach agreement on emission commitments should not be underestimated.

Third, the Kyoto Protocol has the appearance of fairness, in that it focuses on the wealthiest countries and those responsible for a dominant share of the current stock of anthropogenic GHGs in the atmosphere. This is consistent with the principle enunciated in the UNFCCC of "common but differentiated responsibilities and respective capabilities."

Fourth and finally, the fact that the Kyoto Protocol was signed by more than 180 countries and subsequently ratified by a sufficient number of Annex I countries for it to come into force speaks to the political viability of the agreement, if not to the feasibility of all countries actually achieving their targets.

In the realm of public policy, as in our everyday lives, we frequently learn more from our mistakes or failures than from our successes. So, too, in the case of the Kyoto Protocol. Therefore, we also examine some key weaknesses of the Protocol and explore what potentially valuable lessons they may hold for the path forward.

First, it is well known that some of the world's leading GHG emitters are

[4] We use Annex I and Annex B interchangeably to represent those industrialized countries that have commitments under the Kyoto Protocol, though we recognize that a few countries are included in one Annex but not the other.

not constrained by the Kyoto Protocol. The United States – until recently the country with the largest share of global emissions – has not ratified and is unlikely to ratify the agreement. Also, some of the largest and most rapidly growing economies in the developing world do not have emission targets under the agreement. Importantly, China, India, Brazil, South Africa, Indonesia, Korea, and Mexico are not listed in Annex B of the Kyoto agreement. Rapid rates of economic growth in these countries have produced rapid rates of growth in energy use, and hence carbon dioxide (CO_2) emissions. Together with continued deforestation in tropical countries, the result is that the developing world has overtaken the industrialized world in total GHG emissions. China's industrial CO_2 emissions have already surpassed those of the United States; moreover, China's emissions are expected to continue growing much faster than US emissions for the foreseeable future (Blanford, *et al.*).[5]

These realities raise the possibility that the Kyoto Protocol is not as fair as originally intended, especially given how dramatically the world has changed since the UNFCCC divided countries into two categories in 1992. For example, approximately fifty non-Annex I countries – that is, developing countries and some others – now have higher per capita incomes than the poorest of the Annex I countries with commitments under the Kyoto Protocol. Likewise, forty non-Annex I countries ranked higher on the Human Development Index in 2007 than the lowest ranked Annex I country.

A second weakness of the Kyoto Protocol is associated with the relatively small number of countries being asked to take action. This narrow but deep approach may have been well intended, but one of its effects will be to drive up the costs of producing carbon-intensive goods and services within the coalition of countries taking action. (Indeed, increasing the cost of carbon-intensive activities is the intention of the Protocol and is fully appropriate as a means to create incentives for reducing emissions.) Through the forces of international trade, however, this approach also leads to greater comparative advantage in the production of carbon-intensive goods and services for countries that do not have binding emissions targets under the agreement. The result can be a shift in production and emissions from participating nations to non-participating nations – a phenomenon known as emission "leakage." Since leakage implies a shift of industrial activity and associated

---

[5] Citations without a year refer to the author's work in the Harvard Project on International Climate Agreements, which is described in a brief summary in Appendix 1. We also refer to the Foreword in this manner ("Wirth"). Where articles or books outside this volume are referenced, the usual citation with a year is provided, with the full reference provided in the list at the end of Chapter 2.

economic benefits to emerging economies, there is an additional incentive for non-participants to free-ride on the efforts of those countries that are committed to mitigating their emissions through the Protocol's narrow but deep approach.

This leakage will not be one-for-one (in the sense that increased emissions in non-Annex I countries would be expected to fully negate emission reductions in Annex I countries), but it will reduce the cost-effectiveness and environmental performance of the agreement and, perhaps worst of all, push developing countries onto a more carbon-intensive growth path than they would otherwise have taken, rendering it more difficult for these countries to join the agreement later.

A third concern about the Kyoto Protocol centers on the nature of its emission trading elements. The provision in Article 17 for international emission trading is unlikely to be effective (Hahn and Stavins 1999). The entire theory behind the claim that a cap-and-trade system is likely to be cost-effective depends upon the participants being cost-minimizing entities. In the case of private-sector firms, this is a sensible assumption, because if firms do not seek – and indeed succeed in – minimizing their costs, they will eventually disappear, given the competitive forces of the market. But nation-states can hardly be thought of as simple cost-minimizers – many other objectives affect their decision-making. Furthermore, even if nation-states sought to minimize costs, they do not have sufficient information about marginal abatement costs at the multitude of sources within their borders to carry out cost-effective trades with other countries.

There is also concern regarding the CDM. This is not a cap-and-trade mechanism, but rather an emission-reduction-credit system. That is, when an individual project results in emissions below what they would have been in the absence of the project, a credit – which may be sold to a source within a cap-and-trade system – is generated. This approach creates a challenge: comparing actual emissions with what they would have been otherwise. The baseline – what would have happened had the project not been implemented – is unobserved and fundamentally unobservable. In fact, there is a natural tendency, because of economic incentives, to claim credits precisely for those projects that are most profitable, and that hence would have been most likely to go forward even without the promise of credits. This so-called "additionality problem" is a serious issue. There are ways to address it through future restructuring and reform of the CDM; we examine some of these options in several parts of this *Summary for Policymakers*.

Fourth, the Kyoto Protocol, with its five-year time horizon (2008 to 2012), represents a relatively short-term approach for what is fundamentally a long-term problem. GHGs have residence times in the atmosphere of decades to centuries. Furthermore, to encourage the magnitude of technological change that will be required to meaningfully address the threat of climate change, it will be necessary to send long-term signals to the private market that stimulate sustained investment and technology innovation (Newell).

Finally, the Kyoto Protocol may not provide sufficient incentives for countries to comply (Barrett). Some countries' emissions have grown so fast since 1990 that it is difficult to imagine those countries being able to undertake the emission mitigation or muster the political will and resources necessary to purchase enough emission allowances or CDM credits from other countries, so as to comply with their targets under the Protocol. For example, Canada's GHG emissions in 2006 exceeded that country's 1990 levels by nearly 55 percent, making it very unlikely that Canada could comply with an emissions target set at 6 percent *below* 1990 levels, averaged over the 2008–2012 commitment period. In short, the enforcement mechanism negotiated for the Kyoto Protocol does not appear to induce policy responses consistent with agreed-upon targets.

Alternative policy architectures for the post-Kyoto period

In our earlier book, *Architectures for Agreement: Addressing Global Climate Change in the Post-Kyoto World*, we characterized potential post-Kyoto international policy architectures as falling within three principal categories: targets and timetables, harmonized national policies, and coordinated and unilateral national policies (Aldy and Stavins 2007). The policy architectures that have subsequently been examined as part of the Harvard Project on International Climate Agreements – while falling within the same three categories – move substantially beyond what was articulated in our 2007 book. Nevertheless, an overview of international policy architectures through the lens of these three categories, together with some concrete examples, is helpful.

The first category – targets and timetables – is the most familiar. At its heart is a centralized international agreement, top-down in form. This is the basic architecture underlying the Kyoto Protocol: essentially country-level quantitative emission targets established over specified time frames. An example of an approach that would be within this realm of targets and timetables, but would address some of the perceived deficiencies of the Kyoto Protocol, would be a regime that established emission targets based on formulas rather than specified fixed quantities (see Frankel, "Formulas"). In lieu of *ad hoc* negotiations over emission caps, this formula approach would establish principles that could be translated into quantitative metrics for determining emission obligations. These formulas could be structured to have some of the appealing properties of indexed growth targets: setting targets as a function of a country's gross domestic product (GDP) per capita, for example. As countries became wealthier, their targets would become more stringent.[6] Conversely, when and if countries faced difficult economic periods, the stringency of their targets would be automatically reduced.

Such an approach does not divide the world simply into two categories of countries, as in the Kyoto Protocol. Rather, it allows for a continuous differentiation among the countries of the world while including all of them. In this way it reduces – if not eliminates – problems of emission leakage, yet still addresses the key criterion of distributional equity and does so in a more careful, sophisticated manner.

The second category – harmonized domestic policies – focuses more on national policy actions than on goals and is less centralized than the first set of approaches. In this case, countries agree on similar domestic policies. This reflects the view that national governments have much more control over their countries' policies than over their emissions. One example is a set of harmonized national carbon taxes (Cooper).[7] With this approach, each participating country sets a domestic tax on the carbon content of fossil fuels, thereby achieving cost-effective control of emissions within its borders. Taxes would be set by nations, and nations would have complete discretion over the revenues they generate. Countries could design their tax policies to be revenue-neutral – for example, by returning the revenues raised to the economy through proportional cuts in other, distortionary taxes, such as those on labor and capital. In order to achieve global cost-effectiveness, carbon taxes would need to be set at the same level in all countries. This would presumably not be acceptable to the poorer countries of the world.

---

[6] Such a mechanism was proposed by Frankel (2007) and is similar to the graduation mechanism proposed by Michaelowa (2007). As developing countries realize growth in per capita income and per capita emissions on par with Annex I countries, they would be expected to take on binding emission targets. In Appendix 1, Frankel ("Formulas"); Ellerman; Karp and Zhao; and Cao provide examples of the targets-and-timetables approach.

[7] McKibbin and Wilcoxen (2007) advance the idea of parallel, unlinked domestic cap-and-trade programs as a way to move forward in international climate policy. They recommend a harmonized safety-valve price mechanism in their domestic cap-and-trade programs. In Appendix 1, Cooper; Jaffe and Stavins; and Sawa provide examples of harmonized domestic policies.

Therefore, significant side deals would most likely need to accompany such a system of harmonized carbon taxes to make it distributionally equitable and hence politically feasible. This could take the form of large financial transfers through side payments from the industrialized world to the developing world, or agreements in the trade or development agenda that effectively compensate developing countries for implementing carbon taxes.

The third and final category that we have used to classify potential post-Kyoto climate policy architectures is coordinated and unilateral national policies. This category includes the least centralized approaches that we have considered – essentially bottom-up policies that rely on domestic politics to drive incentives for participation and compliance (Pizer 2007).[8] Although these approaches are the least centralized, they should not be thought of as necessarily the least effective. One example of a bottom-up approach – linking independent national and regional tradable permit systems – may already be evolving (see Jaffe and Stavins).

The Bali road map and the path ahead

At the December 2007 UN-sponsored climate change talks in Bali, Indonesia (COP 13), the international community reached agreement on the Bali Action Plan, a two-year road map to guide the negotiation of a framework that builds on and succeeds the Kyoto Protocol. This road map identifies many important issues that merit consideration and resolution in the design of an international climate policy architecture. While the Bali Action Plan is intended to yield an international framework at the 2009 climate change talks in Copenhagen, Denmark (COP 15), the road map also provides something of a framework for the international climate policy debate – and thus for actions undertaken domestically by participating countries – for some years beyond the Copenhagen meetings.

The research program pursued by the Harvard Project on International Climate Agreements addresses key issues in the Bali road map with the aim of informing the design and evaluation of various policies that would be included in the next international climate regime. Specifically, Harvard Project research teams have brought their scholarship to bear on each of the five major elements of the Bali Action Plan: a long-term global climate policy goal, emission mitigation, adaptation, technology transfer, and financing.

---

[8] In Appendix 1, Jaffe and Stavins, and Barrett describe examples of the third type of architecture: bottom-up, coordinated and unilateral national policies.

The Bali road map calls for a "shared vision for long-term cooperative action" that would include "a long-term global goal for emission reductions" as a means to implement the ultimate objective of the UNFCCC. The issue of setting long-term goals has received considerable attention from policymakers around the world. While we recognize that national leaders, rather than scholars, will ultimately decide on a long-term global climate policy goal, our work can still inform the identification and review of various long-term emission objectives. The research undertaken for this project and in writing *Architectures for Agreement* identifies a variety of means for constructing a long-term international climate policy architecture – for example, Bosetti, *et al.* evaluate the long-term GHG concentration and temperature implications of a half dozen approaches to climate policy. Additional analyses highlight the challenge of achieving long-term stabilization targets with incomplete participation (Jacoby, *et al.*; Blanford, *et al.*) as well as the need to improve the technology options available for achieving ambitious long-term emission-reduction goals (Clarke, *et al.*).

The role of emission mitigation continues to be central in international climate change negotiations. The Bali Action Plan calls for "mitigation commitments or actions" by developed countries and "mitigation actions" by developing countries, the latter with support for capacity-building and technology transfer from developed countries. In both cases, mitigation efforts should be "measurable, reportable, and verifiable," a requirement that is addressed by Project research aimed at evaluating various kinds of metrics for assessing mitigation activities (Fischer and Morgenstern) and at describing a surveillance institution that can independently review the comparability of effort among participating countries.

The Bali road map provides guidance for these efforts by identifying several specific forms of mitigation, including reducing deforestation and emissions from changes in land use, an issue investigated by Plantinga and Richards. Sectoral approaches to mitigating emissions also receive attention in the Bali road map; accordingly, Sawa and Barrett, among others, explore the prospects and pitfalls of a sector-specific approach. Finally, the negotiators in Bali also agreed on the general proposition that market-based approaches should be pursued – an issue that receives attention in many contributions to this project (Agarwala; Cooper; Ellerman; Frankel, "Formulas"; Jaffe and Stavins; Karp and Zhao; and Keohane and Raustiala).

The Kyoto Protocol only mentions the word "adaptation" twice. In contrast, the Bali road map elevates the importance of this issue. Several contributors to the Project recognize the need to effectively integrate climate change and economic development in the design of future climate change policy (including Cao; Hall, *et al.*; Somanathan; and Victor). For example, Barrett argues that efforts to transfer resources and facilitate the development of capacity in developing countries should play an important role in the next climate agreement. Newell points out that efforts to promote technological innovation can address adaptation needs, while also identifying new ways to lower the cost of emission mitigation. Others maintain that promoting economic development, diversifying economic activity, and improving economic resilience – especially in agriculture – should guide climate change policy for the least developed countries.

The Bali road map also focuses on the need to enhance technology transfer to developing countries. Given the rapid growth of emissions in these countries, technology transfer is needed to promote a more climate-friendly trajectory for economic development. The Harvard Project has explored potential reforms of the CDM that would focus on moving more technologies to developing countries (Teng, *et al.*); it has also examined options for the design of clean technology funds oriented to developing countries (Hall, *et al.*; Keeler and Thompson). Of course, the success of technology transfer will depend on the development of new technologies – an issue addressed by Newell in exploring the potential for policy to induce more innovation on climate-friendly technologies. Along all of these dimensions of action – mitigation, adaptation, and technology transfer – the Project has assessed opportunities to finance a serious and sufficient climate policy program as called for in the Bali Action Plan.

Finally, the Harvard Project has also advanced research on important issues that, while not identified in the Bali road map, are critical to the design of a successful, post-2012 international climate policy architecture. These include analysis of the equity implications of international climate agreements (Posner and Sunstein); possible means for promoting compliance with internationally-negotiated commitments (Keohane and Raustiala); avenues for structuring a dynamic, robust series of negotiations that can facilitate broad participation and agreement (Harstad); and trade-climate interactions that could enhance an international climate policy agreement (Frankel, "Trade").

Implementing architectures for agreement

In Appendix 1, we provide three-page descriptions of the Harvard Project's diverse set of research initiatives within six sections: (1) alternative international policy architectures; (2) negotiation, assessment, and compliance; (3) the role and means of technology transfer; (4) global climate policy and international trade; (5) economic development, adaptation, and deforestation; and (6) modeling analyses of the impacts of alternative allocations of responsibility.[9]

Alternative international policy architectures

Each of the seven research initiatives that make up Section I of Appendix 1 proposes and assesses a specific post-Kyoto international policy architecture. Jeffrey Frankel of Harvard University proposes "Specific Formulas and Emission Targets for All Countries in All Decades," building on the foundation of the Kyoto Protocol, but strengthening it in important ways. Frankel's approach attempts to solve the most serious deficiencies of Kyoto: the absence of long-term targets, the non-participation of the United States and key developing countries, and the lack of motivation for countries to abide by their commitments. Frankel's plan – which reflects political as well as scientific and economic considerations – uses formulas to set emissions caps for all countries through the year 2100. The methodology is designed to yield caps that give every country reason to feel that it is only doing its fair share, and it is flexible enough that it can accommodate major changes in circumstances during the course of the century.

Denny Ellerman of the Massachusetts Institute of Technology posits that the European Union Emission Trading Scheme (EU ETS) can serve as a prototype for a global policy architecture. Ellerman draws on the first four years of experience with the EU ETS to develop insights about the challenges that can be expected to emerge in a broader program and suggest potential solutions. Interestingly, the problems that are often seen as most intractable for a global trading system – institutional readiness and public acceptance – have not appeared in Europe. Rather, Ellerman finds the greater challenges may lie in developing an effective centralized authority, devising side benefits to encourage participation, and dealing with the interrelated issues of harmonization, differentiation, and stringency. The EU ETS is not perfect, nor does it provide a perfect prototype for a global system, which would surely diverge

---

[9] A total of twenty-seven chapter summaries are included in Appendix I. Seven propose complete international policy architectures (section 1 above); fourteen examine key design issues and elements (sections 2–5); five provide quantitative modeling of alternative policy architectures or allocations of responsibility (section 6); and the last summarizes Richard Schmalensee's Epilogue to *Post-Kyoto International Climate Policy: Implementing Architectures for Agreement* (Aldy and Stavins 2009).

in important respects from the European model. Nevertheless, Ellerman concludes that the EU example is likely to continue to be highly instructive as policymakers consider the larger and more difficult tasks that lie ahead.

The next research initiative continues the focus on tradable permit systems, but does so in the context of a potential decentralized, bottom-up global climate policy architecture. Judson Jaffe of Analysis Group and Robert Stavins of Harvard assess "Linkage of Tradable Permit Systems in International Climate Policy Architecture." The authors note that tradable permit systems are emerging as a preferred policy tool for reducing GHG emissions in many countries around the world. Because linking systems can reduce compliance costs and improve market liquidity, there is great interest in doing so. Jaffe and Stavins examine the benefits and concerns associated with linkage and analyze the near-term and long-term roles that linkage may play in a future international climate policy architecture. They find that in the near term, indirect linkages of cap-and-trade systems via a common emission-reduction-credit system could achieve meaningful cost savings and risk-diversification without the need for much harmonization among systems. In the longer term, international negotiations could establish shared environmental and economic expectations that could serve as the basis for a broad set of direct links among cap-and-trade systems.

Next, we move from global policy architectures based on tradable permit systems to a distinctly different approach, namely a system of harmonized domestic carbon taxes. In "The Case for Charges on Greenhouse Gas Emissions," Richard Cooper of Harvard proposes a world-wide tax on emissions of GHGs from all sources. This approach is premised on the notion that seriously addressing GHG emissions requires a global approach, not one limited to today's rich countries. Levying a charge on CO_2 raises the price of CO_2-emitting activities, including fossil fuel use, and thus is the most direct method of influencing consumer and industrial behavior around the world. The charge would be internationally adjusted from time to time, and each country would collect and keep the revenue it generates. A carbon tax integrated in an existing tax system may be easier to implement, from an institutional perspective, than alternative mitigation policies in some developing countries with weak regulatory bureaucracies and rule of law.

A developing-country perspective is introduced by Ramgopal Agarwala of Research and Information System for Developing Countries in New Delhi, India. Starting from the observation that there has been little progress toward a global consensus on climate policy despite growing awareness of the risks of inaction, Agarwala argues that fundamental differences of perspective between developed and developing countries may impede progress toward a new agreement for quite some time. With this in mind, Agarwala presents an approach intended to reconcile the positions of developing and developed countries. After describing why the Kyoto Protocol satisfies none of the key criteria for a credible global compact, the author posits four fundamentals for a future climate agreement: first, it should set realistic targets designed to stabilize global CO_2 emissions at 2003 levels until 2050 and achieve a 50 percent reduction by 2100; second, it should set appropriate carbon prices by eliminating subsidies to emitters (particularly energy subsidies) and establishing a carbon tax; third, it should support the development and dissemination of carbon-saving technologies; and fourth, it should be negotiated within the UN, but should be implemented using institutions such as the International Monetary Fund and World Bank.

Yet another approach to global climate policy is proposed by Akihiro Sawa of the 21st Century Public Policy Research Institute, in Tokyo. He describes "Sectoral Approaches to a Post-Kyoto International Climate Policy Framework." A number of authors and policymakers from industrialized and developing countries have proposed sectoral approaches to a future international agreement. Though there is significant variation in the details, most of these proposals would determine overall emission targets by estimating and aggregating sector-level reduction potentials based on a technology analysis. This is unlike the Kyoto Protocol, in which economy-wide emission commitments are negotiated from the top down. Sawa reviews the pros and cons of sectoral approaches and proposes a specific example for the post-2012 period. He concludes that a sectoral approach may help solve some of the problems of the Kyoto Protocol, but that some issues – including lower cost-effectiveness compared to an economy-wide approach, the difficulty of collecting the data needed to make a technology-based assessment of reduction potential, and the complexity of sector-level negotiations – remain unresolved.

Finally, Scott Barrett of Columbia University offers an approach that departs dramatically from the tradition of the Kyoto Protocol, namely "A Portfolio System of Climate Treaties." Rather than attempting to address all sectors and all types of GHGs under one unified regime, Barrett argues for a system of linked international agreements that separately address different sectors and gases, as well as key issues (such as adaptation and technology research and development) and last-resort remedies (such as geoengineering). Barrett concludes that his proposed multi-track climate treaty system is not perfect, but could nevertheless offer important advantages over the current approach. In particular, by avoiding the enforcement problems of an aggregate approach and by taking a broader view of risk reduction, the portfolio approach could provide a more effective and flexible response to the long-term global challenge posed by climate change.

Negotiation, assessment, and compliance

The other research initiatives of the Project and the remainder of Appendix 1 focus on specific issues of design that are important, no matter the climate policy architecture that is ultimately chosen. In fact, the ideas on various design elements could be aggregated to serve as the basis for an international agreement. Alternatively, some of these ideas complement the architectures described above and could be integrated with these core proposals. Section II features four analyses that examine three closely related topics: the negotiation process; how to assess commitments and compliance; and how to think about distributional equity and fairness.

Bård Harstad of Northwestern University describes "How to Negotiate and Update Climate Agreements," starting from the premise that the outcome of negotiations depends on the bargaining rules. Drawing on a game theoretic analysis, Harstad proposes several bargaining rules that would facilitate agreement on a post-2012 climate treaty: first, harmonization or formulas should be used to calculate national obligations and contributions; second, a future climate treaty should have a long time horizon; third, the treaty should specify the default outcome if the (re)negotiation process breaks down, and this default outcome should be an ambitious agreement; fourth, investments in research and development (R&D), or trade in abatement technology, should be subsidized internationally; fifth, unanimity requirements – where they exist – should be replaced by a majority or a super-majority rule when it comes to treaty amendments; sixth, linkage to international trade agreements makes each of the rules more credible and efficient; and seventh, a "minimum participation rule" can discourage free-riding.

Developing effective strategies to address climate change will require collective effort on the part of many countries over an extended time horizon and across a range of activities. Therefore, a key challenge for the international community will be to compare and judge different national commitments. Carolyn Fischer and Richard Morgenstern of Resources for the Future take on this topic in "Metrics for Evaluating Policy Commitments in a Fragmented World: The Challenges of Equity and Integrity." Because diverse actions by different nations will be an unavoidable part of future climate policy, it will be critical in international negotiations to have some means of talking in a coherent and broadly accepted manner about what individual nations are doing to help reduce climate risk. Various metrics for evaluating individual nations' policy commitments and performance are considered by Fischer and Morgenstern, who conclude that no single metric can adequately address the complex issues of equity and the integrity of climate change mitigation measures. Rather, a set of metrics will inevitably be required.

Clearly, climate change raises difficult issues of justice, particularly with respect to the distribution of burdens and benefits among poor and wealthy nations. Eric Posner of the University of Chicago and Cass Sunstein of Harvard examine this important topic. In "Justice and Climate Change," these authors focus on the narrower question of how to allocate GHG emission rights within a future international cap-and-trade system. However, the questions they address apply equally to a variety of other mechanisms for allocating cost burdens internationally. They identify shortcomings in an approach that is often advanced on fairness grounds – a per capita allocation in which emissions permits are distributed to nations on the basis of population. Although Posner and Sunstein acknowledge that allocations based on population or on redistributing wealth are generally more equitable than allocations that award permits on the basis of current emissions, they maintain that a per capita allocation would not – in practice – satisfy objectives of fairness and welfare redistribution. Rather, if the goal is a more equal distribution of wealth, an approach that is openly redistributive is better than a per capita allocation.

Ultimately, an international climate agreement will be of no value without sufficient participation and compliance by signatories. This is one of the lessons of the Kyoto Protocol. Robert Keohane of Princeton University and Kal Raustiala of the University of California at Los Angeles begin with the proposition that a successful climate change regime must secure sufficient participation, achieve agreement on meaningful rules, and establish mechanisms for compliance. Moreover, it must do so in a political environment of sovereign states with differing preferences and capabilities. In "Toward a Post-Kyoto Climate Change Architecture: A Political Analysis," Keohane and Raustiala address the trade-off between participation and stringency by proposing an "economy of esteem for climate change," in which participation is encouraged by a system of prizes for politicians who take leadership on this issue. They argue that, contrary to provisions in the Kyoto Protocol, only a system of buyer liability (rather than seller or hybrid liability) in an international permit trading regime is consistent with existing political realities and will promote compliance. Drawing analogies to international bond markets, they propose a system of buyer liability that would endog-enously generate market arrangements, such as rating agencies and fluc-tuations in the price of emissions permits according to perceived risk. These features would in turn create incentives for compliance without resorting to ineffective inter-state punishments.

The role and means of technology transfer

Achieving long-term climate change policy goals will require dramatic progress in the innovation and deployment of energy-efficient and low-carbon technologies (Aldy and Stavins 2008c). Policies that directly facilitate technology innovation and diffusion will therefore need to play a central role, alongside policies targeted directly at reducing emissions. These are addressed in Section III.

Richard Newell of Duke University takes a broad perspective, proposing a portfolio of "International Climate Technology Strategies" within the context of international agreements and institutions for climate, energy, trade, devel-opment, and intellectual property. First, Newell notes that long-term national commitments and policies for emission mitigation are crucial to providing the necessary private-sector incentives for technology development and transfer. Financial assistance to developing countries for technology transfer and capacity building will also be necessary. Tariff and non-tariff barriers to the transfer of climate-friendly technologies can be reduced through a World Trade Organization (WTO) agreement on trade in environmental goods and services. To support the upstream supply and transfer of technology innovations internationally, Newell proposes strategies to increase and more effectively coordinate public funding of R&D, as well as strategies to resolve impediments to knowledge transfer. The result is a portfolio of strategies for reducing barriers and increasing incentives for innovation across interna-tional agreements and institutions.

An agenda focused on technology transfer is laid out by Andrew Keeler and Alexander Thompson of Ohio State University, "Mitigation through Resource Transfers to Developing Countries: Expanding Greenhouse Gas Offsets." Keeler and Thompson propose a more expansive approach to offsets that would meet the different objectives of industrialized and developing countries, while providing substantial support for long-term investments and policy changes to reduce GHG emissions in the developing world. Their approach consists of five elements: (1) change the criteria for offsets from "real, verifiable, and permanent reductions" to "actions that create real progress in developing countries toward mitigation and adaptation"; (2) make a significant share of industrialized country commitments achievable through offset payments to developing countries; (3) sell a portion of offset credits up front and put the proceeds in a fund to make investments in projects in the developing world; (4) focus international negotiations on guidelines for an international offsets program; and (5) delegate tasks to new and existing institutions for the purpose of managing the offsets program.

In "Possible Development of a Technology Clean Development Mechanism in the Post-2012 Regime," Fei Teng, Wenying Chen, and Jiankun He of Tsinghua University in Beijing offer a proposal that is parallel to the Keeler and Thompson proposal, but that fits within the context of an enhanced CDM. Starting from the premise that it will be essential to transfer climate-friendly technologies from developed to developing countries, the authors propose an enhanced CDM regime with a specific emphasis on technology transfer. This enhanced regime would have three features: first, technology transfer must be identified as a goal before any activities are approved and implemented; second, only projects that use technology transferred under the program can receive credit for emissions reductions; and third, credits would be shared by the technology provider or by the government of the host country if the technology provider or host-country government support or enable the transfer, as well as offer discounted or even free licensing.

Global climate policy and international trade

Global efforts to address climate change could be on a collision course with global efforts through the WTO to reduce barriers to trade. This is the theme of Section IV. Such a collision would be terrible news – both for free trade and for climate protection. In "Global Environment and Trade Policy," Jeffrey Frankel of Harvard University first examines the broad question of whether environmental goals in general are threatened by free trade and the WTO, before turning to the narrower question of whether trade policies that could be included in various national efforts to address climate change are likely to come into conflict with WTO rules. Frankel notes that future national-level policies to address climate change are likely to include provisions that target carbon-intensive products from countries deemed to be making inadequate efforts. These provisions need not violate sensible trade principles and WTO rules, but there is a danger that in practice they will. Frankel describes the characteristics of future national policies that would likely conflict with WTO rules and could provide cover for protectionism – he also describes the characteristics of future national policies that likely would be WTO-compatible. Frankel concludes that in the long term, a multilateral regime is needed to guide the development of trade measures intended to address concerns about leakage and competitiveness in a world where nations have different levels of commitment to GHG mitigation.

In "A Proposal for the Design of the Successor to the Kyoto Protocol," Larry Karp of the University of California at Berkeley and Jinhua Zhao of Michigan State University examine how international trade mechanisms can be made part of a future climate agreement. In their proposal, nations with mandatory emissions ceilings would have the option to reduce their abatement commitments in exchange for either paying a monetary fine or accepting trade sanctions imposed by other signatory nations. In addition to the potential use of trade sanctions, trade reforms could be used to achieve climate-related objectives. Specifically, these authors support the use of carefully circumscribed border tax adjustments to protect against leakage. They maintain that such adjustments – if thoughtfully and carefully applied – can create effective incentives for countries to participate in a future agreement.

Economic development, adaptation, and deforestation

Developing countries have a key role to play in efforts to address climate change – both because they could be strongly affected by future damages and because they account for an increasing share of global emissions. For this reason, the links between international climate policy and economic development are enormously important. Policies to facilitate adaptation and reduce the rate of deforestation, in particular, are critical for developing countries. Because of the great significance of this set of issues in the post-Kyoto international climate policy debate, we feature five research initiatives on economic development, adaptation, and deforestation in Section V.

Jing Cao of Tsinghua University provides a Chinese perspective on "Reconciling Human Development and Climate Protection." Describing an approach that shares much with the proposal by Jeffrey Frankel ("Formulas"), Cao seeks to offer a fair and efficient policy architecture for the post-2012 era, with the hope of breaking through what she characterizes as the current political impasse between developed and developing countries. Cao's proposed approach engages developing countries gradually, through four stages: in the first stage, all countries agree on a path of future global emissions that leads to an acceptable long-term stabilization goal; in the second stage, developing countries focus on "no regrets" mitigation options; in stage three, developing countries take on moderate emissions targets; and in the final stage, all countries agree to binding emissions targets.

E. Somanathan of the Indian Statistical Institute in New Delhi recognizes that an effective solution to the climate change problem will require the cooperation of developing countries ("What Do We Expect from an International Climate Agreement? A Perspective from a Low-Income Country"). However, he argues that it is neither feasible nor desirable to pursue near-term GHG reductions within these countries or emissions trading between developed and developing countries. Arguing that technology improvements are needed to give all countries, including developing countries, a realistic opportunity to cost-effectively cut their CO_2 emissions, Somanathan maintains that a post-2012 international climate agreement should focus on creating incentives for research and development to advance new climate-friendly technologies. Indeed, he indicates that an international agreement involving developing countries should confine itself to promoting technical cooperation.

David Victor of Stanford University takes a different approach to engaging developing countries. He proposes "Climate Accession Deals: New Strategies for Taming Growth of Greenhouse Gases in Developing Countries." This approach builds on two premises: first, that developing nations value economic growth far more than they value future global environmental conditions, and second, that many governments of developing nations lack the administrative ability to control emissions. With Victor's proposal, climate accession deals would be negotiated on a country-by-country basis, with an individual accession deal consisting of a set of policies that are tailored to gain maximum leverage on a single developing country's emissions, while still being aligned with its interests and capabilities. Industrialized countries would support each accession deal by providing specific benefits, such as financial resources, technology, administrative training, or security guarantees. According to Victor, accession deals could have several advantages: first, they would be anchored in host countries' interests and capabilities; second, they could yield a significant degree of leverage while minimizing external investment; third, they would engage private enterprise and government ministries other than environmental and foreign affairs ministries; and fourth, accession deals would be replicable and scalable.

Daniel Hall of Resources for the Future, Michael Levi of the Council on Foreign Relations, William Pizer of Resources for the Future, and Takahiro Ueno of the Central Research Institute of the Electric Power Industry in Tokyo offer a broad approach to "Policies for Developing Country Engagement."

These authors maintain that because no single approach offers a sure path to success for securing developing-country participation, a variety of strategies – including policy reforms, financing approaches, and diplomatic avenues – must be pursued in parallel. In their view, post-Kyoto international climate negotiations are likely to focus on a "grand bargain" with developing countries, offering some form of commitments in exchange for further emission reductions and increased financing from developed countries. Developing country commitments could take the form of domestic policy reforms, sectoral targets, or even economy-wide limits (for higher-income developing countries). These authors conclude that forging a new climate agreement that reduces global emissions and provides support to poor countries will be very difficult, but without it there is virtually no chance of stabilizing GHG concentrations at an acceptable level.

Changes to forests worldwide can have enormous impacts on the global carbon cycle. Because of this, Andrew Plantinga of Oregon State University and Kenneth Richards of Indiana University argue – as do increasing numbers of scholars and policymakers – that forest carbon management ought to be an element of the next international agreement on climate change. In "International Forest Carbon Sequestration in a Post-Kyoto Agreement" they propose a "national inventory" approach, in which nations receive credits or debits for changes in forest cover relative to a measured baseline. Nations would conduct periodic inventories of their forest carbon stock, and the measured stock would be compared with a pre-negotiated baseline to determine offset credits that could be redeemed, or debits that must be covered, in a tradable permit market. With this approach, national governments, rather than project developers, would pursue carbon sequestration activities through the implementation of domestic policies.

Modeling impacts of alternative allocations of responsibility

Clearly, negotiations on a post-Kyoto international climate regime will be driven in large part by the perspectives of individual countries that are primarily concerned about the impacts of any future agreement on their own economies and societies. Just as no single individual or institution has cornered the market on wisdom regarding the best architecture for a post-Kyoto climate policy, so too has no single economic model captured all dimensions and concerns regarding the consequences of alternative allocations of responsibility. Hence, Section VI includes five sets of modeling results obtained by research teams on three continents.

In "A Quantitative and Comparative Assessment of Architectures for Agreement," Valentina Bosetti, Carlo Carraro, Alessandra Sgobbi, and Massimo Tavoni, all of the Fondazione Eni Enrico Mattei in Italy, provide a comparison of eight prominent options: global cap-and-trade with redistribution; global tax recycled domestically; reducing emissions from deforestation and degradation; climate clubs; burden sharing; graduation; dynamic targets; and R&D and technology development. They assess these architectures in terms of four criteria: economic efficiency, environmental effectiveness, distributional implications, and political acceptability, as measured by feasibility and enforceability. The authors conclude, first, that a stabilization target of 450 parts per million (ppm) for the atmospheric concentration of CO_2 only (550 ppm for all GHGs in CO_2-equivalent terms) is hardly achievable. However, a strategy of progressive commitments – in which consensus is reached on future binding targets for developing countries, but developed countries take action first – can achieve CO_2 stabilization very close to 450 ppm. Second, an extended – possibly global – carbon market, even without global commitments to reduce emissions, will help to reduce costs, as will the inclusion of non-CO_2 gases and credits for avoided deforestation. However, a basic trade-off between economic impact and environmental protection remains.

Henry Jacoby, Mustafa Babiker, Sergey Paltsev, and John Reilly of the Massachusetts Institute of Technology write about "Sharing the Burden of GHG Reductions." They use the MIT Emissions Prediction and Policy Analysis (EPPA) model to estimate the welfare and financial implications of various cost and emission-reduction outcomes. They find that a target of reducing global emissions by 50 percent by 2050, while it can be done in a way that meets reasonable equity targets, is extremely ambitious and would require large financial transfers from developed to developing countries. The authors conclude that the combination of aggressive targets with expectations of incentives and compensation for the developing countries may not reflect sufficient regard for the difficulty of finding a mutually acceptable way to share the economic burden.

Leon Clarke, Kate Calvin, Jae Edmonds, Page Kyle, and Marshall Wise of the Pacific Northwest National Laboratory focus on "Technology and International Climate Policy" to explore interactions between two of the key drivers that determine emissions reductions – technology availability and performance, on the one hand, and international policy architectures, on the other. Four main findings emerge from this analysis: first, technology is more important to reducing the costs of emissions mitigation when international policy structures deviate from full participation; second, near-term carbon prices are inexorably tied to the expected long-term availability of technology; third, the choice of a policy architecture has a larger impact on the distribution of mitigation actions than on the global emissions pathway; and fourth, more rapid technology improvements reduce the relative influence of policy architecture.

Geoffrey Blanford and Richard Richels of the Electric Power Research Institute and Thomas Rutherford of the Swiss Federal Institute of Technology in Zurich examine "Revised Emissions Growth Projections for China: Why Post-Kyoto Climate Policy Must Look East." The authors note that continued growth in developing-country emissions could put stabilization targets effectively out of reach within the next 10 to 20 years, regardless of what wealthier countries do. They suggest that a CO_2 stabilization target of 450 ppm is probably no longer realistic, and a target of 550 ppm now appears as challenging as 450 ppm appeared just a few years ago. However, stabilization at 550 ppm may still be feasible if developed countries undertake immediate reductions and developing countries follow a "graduated accession" scenario, in which China and other mid-income countries (for example, Korea, Brazil, Mexico, and South Africa) join global mitigation efforts in 2020, India joins in 2040, and poorer countries delay participation until 2050. On the other hand, their analysis indicates that if developing countries enter into a global regime more gradually – for example, by adopting progressively more stringent targets only as incomes rise – global emissions may continue to grow through 2050, and even the 550 ppm target will begin to look doubtful. These authors conclude that no issue is more urgent for international climate negotiations than that of establishing incentives for timely and meaningful participation by developing countries.

In "Expecting the Unexpected: Macroeconomic Volatility and Climate Policy," Warwick McKibbin of the Australian National University, Adele Morris of the Brookings Institution, and Peter Wilcoxen of Syracuse University focus on a timely concern: namely how a future international climate policy architecture may perform in the presence of unexpected macroeconomic shocks, whether positive shocks from economic growth in developing countries or severe financial distress in the global economy. Their premise is that, in the absence of such unanticipated economic shocks, three regimes are similar – in principle – in their ability to reduce emissions efficiently: global carbon cap-and-trade; globally harmonized domestic carbon taxes; and a hybrid system of national long-term permit trading with a globally-coordinated maximum price for permits in each year (that is, a safety valve).[10] However, these three systems differ in how they would transmit economic disruptions from one economy to another. McKibbin, Morris, and Wilcoxen find that whereas a cap-and-trade regime would be counter-cyclical – in the sense that reduced demand for permits would lead to lower permit prices and thereby dampen cost impacts during an economic slowdown – this approach also fails to capture the opportunity for significant additional low-cost emissions reductions during a global economic downturn.

Synthesis

In an Epilogue, Richard Schmalensee of the Massachusetts Institute of Technology steps back and reflects on the factors that make the global dimensions of climate change so difficult and important to manage, the history of climate policy debates, and the key elements of an emerging international policy architecture. He concludes that the most critical and difficult task before the world's policymakers is to "move toward a policy architecture that can induce the world's poor nations to travel a much more climate-friendly path to prosperity than the one today's rich nations have traveled."

In the following part of this *Summary for Policymakers*, we highlight certain principles that the research teams have identified as being important for the design of a scientifically sound, economically rational, and politically pragmatic post-2012 international climate policy architecture. Real progress will require addressing these principles, which constitute some of the core premises underlying various policy architectures and design elements. We also highlight four international climate policy architectures – each of which has advantages as well as disadvantages – because each is promising in some regards and because each raises important issues for consideration. One is within the category of targets and timetables: formulas for dynamic national targets for all countries. Two are within the category of harmonized domestic policies: a portfolio of international treaties and harmonized national carbon taxes. And one is within the category of coordinated and unilateral national policies: linkage of national and regional tradable permit systems.

Regardless of which overall international policy architecture is ultimately chosen, a number of key design issues stand out as particularly important. And so, in the last section of Chapter 2, we highlight some of the lessons

---

[10] Refer to McKibbin and Wilcoxen (2007) for a detailed description of this type of climate policy architecture.

identified by our twenty-six research teams with regard to five issues and elements for a post-2012 international agreement: burden-sharing, technology transfer, CDM reform, addressing deforestation, and making global climate policy compatible with global trade policy. We infuse all five of these discussions with attention to the relationship between global climate policy and economic development.

The principles, architectures, and design elements proposed and examined in this *Summary for Policymakers* and highlighted in Chapter 2 can serve to illuminate many of the issues facing the international policy community. Our hope is that all those engaged in the ultimate design of climate change policy – from decision-makers and diplomats to leaders in the private sector and civil society – will find it useful in reconciling their diverse interests and moving forward with effective solutions to the enormous, collective challenge posed by global climate change.

Lessons for the International Policy Community

Joseph E. Aldy and Robert N. Stavins

The nations of the world confront a tremendous challenge in designing and implementing an international policy response to the threat of global climate change that is scientifically sound, economically rational, and politically pragmatic. It is broadly acknowledged that the relatively wealthy, developed countries are responsible for a majority of the anthropogenic greenhouse gases (GHGs) that have already accumulated in the atmosphere, but developing countries will emit more GHGs over this century than the currently industrialized nations if no efforts are taken to alter their course of development. The architecture of a robust international climate change policy will need to take into account the many dimensions and consequences of this issue with respect to the environment, the economy, energy, and development.

Drawing upon lessons from experience with the Kyoto Protocol (Schmalensee)[1] and insights from economics, political science, international relations, legal scholarship, and other disciplines, the contributors to the Harvard Project on International Climate Agreements have set forth a range of ideas about how best to construct a post-2012 international climate change policy regime. The targets-and-timetables approach embodied in the Kyoto agreement appears here in proposals advanced by Jeffrey Frankel, Denny Ellerman, Larry Karp and Jinhua Zhao, and Jing Cao. A second category of international climate policy architectures – harmonized domestic policies – is represented in proposals by Scott Barrett, Judson Jaffe and Robert Stavins, Richard Cooper, and Akihiro Sawa. And one proposal by Judson Jaffe and Robert Stavins falls in a third category: decentralized, bottom-up approaches that rely primarily on coordinated, unilateral national policies. Combined with nineteen analyses that focus on specific design issues, these proposals cover virtually the entire spectrum of potential international climate policy architectures.

---

[1] As in Chapter 1 of this *Summary for Policymakers*, all citations to authors refer, unless otherwise specified, to the author's work in Appendix 1 of this report. Where articles or books outside this report are referenced, the usual citation with year of publication is provided.

This chapter of the *Summary for Policymakers* provides a synthesis of this exceptionally diverse set of proposals and analyses. We begin by identifying a set of principles that our research teams have explicitly or implicitly identified as being important for the design of post-2012 international climate policy architecture. We then go on to highlight four potential architectures, each of which is promising in some regards and raises important issues for consideration. Finally, we turn to key design issues in international policy architecture, because regardless of which overall architecture is ultimately chosen, certain key design issues and elements will stand out as particularly important. We conclude with a look at the path ahead.

Principles for an international agreement

A set of core principles emerges from the diverse strands of research reported in this *Summary for Policymakers*. These principles constitute the fundamental premises that underlie various proposed policy architectures and design elements; as such they can provide a reasonable point of departure for ongoing international negotiations.[2]

Climate change is a global-commons problem, and therefore a cooperative approach involving many nations – whether through a single international agreement or some other regime – will be necessary to address it successfully. Because GHGs mix uniformly in the atmosphere, the location of emissions sources has no effect on the location of impacts, which are dispersed worldwide. Hence, it is virtually never in the economic interest of individual nations to take unilateral actions. This classic free-rider problem means that cooperative approaches are necessary (Aldy and Stavins 2008a).

Since sovereign nations cannot be compelled to act against their wishes, successful treaties should create adequate incentives for compliance, along with incentives for participation. Unfortunately, the Kyoto Protocol seems to lack incentives of both types (Barrett; Karp and Zhao; and Keohane and Raustiala).

Since carbon-intensive economies cannot be replicated throughout the world without causing dangerous anthropogenic interference with the global climate, it will be necessary for all countries to move onto much less carbon-intensive growth paths. Even reducing emissions in the currently industrialized world to zero is insufficient (Blanford, *et al.*; Bosetti, *et al.*; Cooper; Hall, *et al.*; and

---

[2] Aldy, Barrett, and Stavins (2003) present six criteria for evaluating potential international climate policy architectures that map closely to most of these principles.

Jacoby, *et al.*). With appropriate negotiating rules (Harstad), more countries can be brought on board. The rapidly emerging middle class in the developing world seeks to emulate lifestyles that are typical of the industrialized world and may be willing to depart from this goal only if the industrialized world itself moves to a lower-carbon path (Agarwala; Schmalensee; and Wirth). Moving beyond the current impasse will require that developed countries achieve meaningful near-term emission reductions, with a clear view to medium- and long-term consequences and goals (Agarwala; Harstad; and Karp and Zhao).

A credible global climate change agreement must be equitable. If past or present high levels of emissions become the basis for all future entitlements, the developing world is unlikely to participate (Agarwala). Developed countries are responsible for more than 50 percent of the accumulated stock of anthropogenic GHGs in the atmosphere today, and their share of near-term global mitigation efforts should reflect this responsibility (Agarwala). In the long term, nations should assume the same or similar burdens on an equalized per capita basis (Agarwala; Cao; and Frankel, "Formulas"). However, if the goal is a more equitable distribution of wealth, approaches based on metrics other than per capita emissions can be better (Jacoby, *et al.*; and Posner and Sunstein). It is also important to recognize and acknowledge that in the short term, developing countries may value their economic growth more than future, global environmental conditions (Victor).

Developing countries face domestic imperatives for economic growth and political development. More and better research is needed to identify policies that promote both mitigation and adaptation, while accommodating development. At the same time, developing countries should not "hide behind the poor" (Agarwala): The burgeoning middle class in the developing world is on a path to exceed the population of developed countries and, as we have already noted, its lifestyle and per capita emissions are similar to those in much of the developed world. While not exclusively a problem of developing countries, tropical forests, in particular, are one important dimension of the larger interplay between development and climate change policy. Because of the enormous impacts that natural and anthropogenic changes in forests have on the global carbon cycle, it is important to provide a meaningful, cost-effective, and equitable approach to promoting forest carbon sequestration in an international agreement (Plantinga and Richards).

A credible global climate change agreement must be cost-effective. That means it should minimize the global welfare loss associated with reducing emissions (Aldy and Stavins 2008b; Ellerman; and Jaffe and Stavins), and also minimize the risks of corruption in meeting targets (Agarwala; Somanathan).

A credible global climate change agreement must bring about significant technological change. Given the magnitude of the problem and the high costs that will be involved, it will be essential to reduce mitigation costs over time through massive technological invention, innovation, diffusion, and utilization (Blanford, *et al.*; Bosetti, *et al.*; Clarke, *et al.*; Newell; Somanathan; Wirth; and Aldy and Stavins 2008c). Rapid technology transfer from the developed to the developing world will be needed (see Hall, *et al.*; Keeler and Thompson; Newell; Somanathan; Teng, *et al.*; and Wirth).

Governments should work through a variety of channels to achieve a credible global climate change agreement that uses multiple ways to mitigate climate change risks. Although a post-2012 agreement under the UNFCCC may be at the core of a post-Kyoto regime, other venues – whether bilateral treaties, or G8+5, or L20 accords – should continue to be explored, as additional agreements and arrangements may be necessary (Hall, *et al.*; and Schmalensee).

An effective global climate change agreement must be consistent with the international trade regime. A global climate agreement can lead to conflicts with international trade law, but it can also be structured to be mutually supportive of global trade objectives (Frankel, "Trade"; and Harstad).

A credible global climate change agreement must be practical, realistic, and verifiable. That means it needs institutional mechanisms for effective implementation (Agarwala). Because tremendous start-up costs are usually incurred in creating new institutions, consideration should be given – whenever appropriate – to maintaining existing institutions, such as the Clean Development Mechanism, and improving them rather than abandoning them (Hall, *et al.*; Karp and Zhao; Keeler and Thompson; and Teng, *et al.*). In addition, it should be recognized that most parts of the industrialized world have signaled their preference for the use of cap-and-trade mechanisms to meet their domestic emissions commitments (Jaffe and Stavins), and it would be *politically* practical to build upon these institutional and policy preferences. Whatever institutions or mechanisms are used to implement policy commitments, they should promote emission abatement consistent with realistic technological innovation to avoid risking costly and ineffective outcomes (Agarwala; Blanford, *et al.*; Bosetti, *et al.*; and Jacoby, *et al.*). The best agreements will be robust in the face of inevitable global economic downturns (McKibbin, *et al.*). Finally, various metrics can be employed to judge the equity and integrity of national commitments, including measures of emissions performance, reductions, or cost (Fischer and Morgenstern). An international surveillance institution could provide credible, third-party assessments of participating countries' efforts.

Promising international climate policy architectures

While we have identified a number of core principles to guide the analysis and frame the proposals presented in this volume, the Harvard Project does not endorse a single approach to international climate policy. This is because we recognize that the decision to adopt a particular architecture is ultimately a political one, which must be reached by the nations of the world through taking into account a complex array of factors. We highlight four potential architectures – each with advantages as well as disadvantages – because each is promising in some regards, raises key issues for consideration, and to a considerable extent is exemplary of the types of architectures discussed in this volume.

One architecture follows a targets and timetables structure, using formulas to set dynamic national emissions targets for all countries. Two fall within the category of harmonized domestic policies: a portfolio of international treaties and harmonized national carbon taxes. The fourth architecture summarized below is based on a set of coordinated, unilateral national policies and involves linking national and regional tradable permit systems.

Targets and timetables: formulas for evolving emission targets for all countries[3]

This targets-and-timetables proposal offers a framework of formulas that yield numerical emissions targets for all countries through the end of this century (Frankel, "Formulas"). National and regional cap-and-trade systems for greenhouse gases would be linked in a way that allows trading across firms and sources (Jaffe and Stavins), not among nations *per se* (as in Article 17 of the Kyoto Protocol). Such a global trading system would be roughly analogous to the system already established in the European Union, where sources rather than nations engage in trading (Ellerman).[4]

---

[3] This proposed architecture was developed by Frankel, supplemented by Aldy and Stavins (2008b), Harstad, Cao, Ellerman, and Jacoby, *et al.* Bosetti, *et al.* provide an economic analysis of this and several other potential architectures.

[4] For an examination of the possible role and design of cap-and-trade and other tradable permit systems as part of an international policy architecture, see Aldy and Stavins (2008b).

The formulas are based on what is possible politically, given that many of the usual science- and economics-based proposals for future emission paths are not dynamically consistent – that is, future governments will not necessarily abide by commitments made by today's leaders. Several researchers have observed that when participants in the policy process discuss climate targets, they typically pay little attention to the difficulty of finding mutually acceptable ways to share the economic burden of emission reductions (Bosetti, *et al.*; Jacoby, *et al.*).

This formula-based architecture is premised on four important political realities. First, the United States may not commit to quantitative emission targets if China and other major developing countries do not commit to quantitative targets at the same time. This reflects concerns about economic competitiveness and carbon leakage. Second, China and other developing countries are unlikely to make sacrifices different in character from those made by richer countries that have gone before them. Third, in the long run, no country can be rewarded for having "ramped up" its emissions well above 1990 levels. Fourth, no country will agree to bear excessive cost. (Harstad adds that use of formulas can render negotiations more efficient.)

The proposal calls for an international agreement to establish a global cap-and-trade system, where emissions caps are set using formulas that assign quantitative emissions limits to countries in every year through 2100. The formula incorporates three elements: a progressivity factor, a latecomer catch-up factor, and a gradual equalization factor. The progressivity factor requires richer countries to make more severe cuts relative to their business-as-usual emissions. The latecomer catch-up factor requires nations that did not agree to binding targets under the Kyoto Protocol to make gradual reductions to account for their additional emissions since 1990. This factor prevents latecomers from being rewarded with higher targets, and is designed to avoid creating incentives for countries to ramp up their emissions before signing on to the agreement. Finally, the gradual equalization factor addresses the complaint that rich countries are responsible for a majority of the accumulated anthropogenic GHGs currently in the atmosphere. In the second half of the century, this factor moves national per capita emissions in the direction of the global average of per capita emissions.[5]

---

[5] This is similar to Cao's "global development rights" (GDR) burden-sharing formula and is consistent with calls for movement toward per capita responsibility by Agarwala. On the other hand, it contrasts with the analyses of Jacoby, *et al.*, and Posner and Sunstein. Under Cao's GDR formula, the lion's share of the abatement burden would fall on the industrialized world in the short term, with developing countries initially accepting a small but increasing share over time, such that, by 2020, fast-growing economies such as China and India would take on significant burdens.

The caps set for rich nations would require them to undertake immediate abatement measures. Developing countries would not bear any cost in the early years, nor would they be expected to make any sacrifice that is different from the sacrifices of industrialized countries, accounting for differences in income. Developing countries would be subject to binding emission targets that would follow their business-as-usual (BAU) emissions in the next several decades.[6] National emission targets for developed and developing countries alike should not cost more than 1 percent of GDP in present value terms, or more than 5 percent of GDP in any given year.

Every country under this proposal is given reason to feel that it is only doing its fair share. Importantly, without a self-reinforcing framework for allocating the abatement burden, announcements of distant future goals may not be credible and so may not have desired effects on investment. The basic architecture of this proposal – a decade-by-decade sequence of emission targets determined by a few principles and formulas – is also flexible enough that it can accommodate major changes in circumstances during the course of the century.

Harmonized domestic policies: a portfolio of international treaties[7]

The second proposal we highlight is for a very different sort of architecture than that of the Kyoto Protocol. Rather than attempting to address all sectors and all types of GHGs under one unified regime, this approach envisions a system of linked international agreements that separately address various sectors and gases; as well as key issues, including adaptation and technology research and development (R&D); plus last-resort remedies, such as geoengineering and air capture of greenhouse gases.

First, nations would negotiate sector-level agreements that would establish global standards for specific sectors or categories of GHG sources. Developing countries would not be exempted from these standards, but would receive financial aid from developed countries to help them comply. Trade sanctions would be available to enforce agreements governing trade-sensitive sectors. Such a sectoral approach could have the advantage that it protects against cross-contamination: if policies designed for a given sector

---

[6] Somanathan would argue against including developing countries in the short term, even with targets equivalent to BAU, as recommended in this proposal. We discuss alternative burden-sharing arrangements below.

[7] This proposed architecture was developed by Barrett and supplemented by Newell on research and development policies, by Sawa on sectoral approaches, and by economic modeling from Bosetti, *et al.*

prove ineffective, their failure need not drag down the entire enterprise. Similar arguments can be made for separate approaches to different types of GHGs.

In general, sectoral approaches in a future climate agreement can offer some advantages (Sawa). First, sectoral approaches could encourage the involvement of a wider range of countries, since incentives could be targeted at specific industries in those countries. Second, sectoral approaches can directly address concerns about international competitiveness and leakage. If industries make cross-border commitments to equitable targets this would presumably mitigate concerns about unfair competition in energy-intensive industries. Third, sectoral approaches could be designed to promote technology development and transfer. It should also be recognized, however, that sectoral approaches have some significant problems (Sawa). First, it may be difficult to negotiate an international agreement using this approach if negotiators are reluctant to accept the large transaction costs associated with collecting information and negotiating at the sector level. Countries that are already participating in emission trading schemes may tend to avoid any approach that creates uncertainty about their existing investments. Second, a sectoral approach would reduce cost-effectiveness relative to an economy-wide cap-and-trade system or emission tax. Finally, it is difficult for a sectoral approach to achieve high levels of environmental effectiveness, because it does not induce mitigation actions by all sectors.

Recognizing the technology challenge implicit in successfully addressing climate change, a second component of this suite of international agreements could focus on research and development. Specifically, it could require participants to adopt a portfolio of strategies for reducing barriers and increasing incentives for innovation in ways that maximize the impact of scarce public resources and effectively engage the capacities of the private sector (Newell).[8] R&D obligations could be linked with emission reduction policies. For example, an agreement could require all new coal-fired power stations to have certain minimum thermal efficiency – and ready capacity to incorporate carbon capture and storage, as the latter becomes technically and financially feasible – with these obligations binding on individual countries as long as the treaty's minimum participation conditions were met. Such an agreement would reduce incentives for free-riding and could directly spur R&D

---

[8] In the section below on key design issues, we focus on technology transfer as a key design issue for any international climate policy architecture. Bosetti, *et al.* analyze the costs and effectiveness of R&D strategies compared with alternative architectures.

investments in areas where countries and firms might otherwise be likely to under-invest.

Third, an international agreement should address adaptation assistance for developing countries. All nations have strong incentives to adapt, but only rich countries have the resources and capabilities to insure against climate change risks. Rich countries may substitute investments in adaptation – the benefits of which can be appropriated locally – for investments in mitigation, the benefits of which are distributed globally. If so, this would leave developing countries even more exposed to climate risks and widen existing disparities. Critical areas for investment include agriculture and tropical medicine. Policy design to leverage such investment can improve developing countries' resilience to climate shocks while facilitating their economic development.

A fourth set of agreements would govern the research, development, and deployment of geoengineering and air capture technologies.[9] Geoengineering could serve as an insurance policy in case refinements in climate science over the next several decades suggest that climate change is much worse than currently believed and that atmospheric concentrations may have already passed important thresholds for triggering abrupt and catastrophic impacts. Geoengineering may turn out to be cheap relative to transforming the fossil-fuel foundation of industrial economies. While no one country can adequately address climate change through emissions abatement, individual nations may be able to implement geoengineering options. The challenge may lie in preventing nations from resorting to it too quickly or over other countries' objections.

This portfolio approach to international agreements could avoid the enforcement problems of a Kyoto-style targets-and-timetables structure, while providing the means to prevent climate change (through standards that lower emissions), become accustomed to climate change (through adaptation), and fix it (through geoengineering). By avoiding the enforcement problems of an aggregate approach and by taking a broader view of risk reduction, the portfolio approach could provide a more effective and flexible response to the long-term challenge posed by climate change.

---

[9] Geoengineering strategies attempt to limit warming by reducing the amount of solar radiation that reaches the Earth's surface – the most commonly discussed approach in this category involves throwing particles into the atmosphere to scatter sunlight. Air capture refers to strategies for removing carbon from the atmosphere. Possible options include fertilizing iron-limited regions of the oceans to stimulate phytoplankton blooms or using a chemical sorbent to directly remove carbon from the air.

Harmonized Domestic Policies: A System of National Carbon Taxes[10]

This architecture consists of harmonized domestic taxes on GHG emissions from all sources. The charge would be internationally adjusted from time to time, and each country would collect and keep the revenues it generates (Cooper). Since decisions to consume goods and services that require the use of fossil fuels are made on a daily basis by more than a billion households and firms around the world, the most effective way to reach all these decision makers is by changing the prices they pay for these goods and services. Levying a charge on CO_2 emissions does that directly.

Carbon taxes could have several advantages over a cap-and-trade system. First, the allocation of valuable emission allowances to domestic firms or residents under a cap-and-trade scheme could foster corruption in some countries. A carbon tax would avoid such problematic transfers. Likewise, a carbon tax minimizes bureaucratic intervention and the necessity for a financial trading infrastructure (Agarwala). Second, a carbon charge would generate significant revenues that could be used to increase government spending, reduce other taxes, or finance climate-relevant research and development – though it should be noted that the same is true of a cap-and-trade system that auctions allowances. Third, a carbon tax may be less objectionable to developing nations than an emission cap because it does not imply a hard constraint on growth (Pan 2007).[11] Fourth, any international climate regime requires some means for evaluating national commitments and performance (Fischer and Morgenstern). A carbon tax system provides a straightforward and useful metric, since the marginal cost of abatement activities is always equivalent to the tax rate itself.

Since several economies, most notably the European Union, have embarked on a cap-and-trade system, Cooper investigates whether cap-and-trade systems and tax systems can co-exist. He concludes that the answer is "yes," provided that several conditions are met. First, allowance prices under the cap-and-trade system should average no less than the internationally agreed

---

[10] This proposed architecture was developed by Cooper and supplemented by Fischer and Morgenstern on measurement issues, McKibbin, *et al.* on a hybrid of this approach, and economic modeling by Bosetti, *et al.*

[11] China's 2007 National Program on Climate Change indicated that any near-term emissions reductions in that country will be accomplished using domestic policies designed to address energy efficiency, renewable and nuclear energy, and energy security. The document also indicated that in the longer term, China might be willing to place a price on carbon emissions using more direct mechanisms such as an emissions tax or cap-and-trade system (Jiang 2008). This policy approach is reinforced in Part III of China's October 2008 White Paper on climate change (Information Office of the State Council 2008).

carbon tax. Second, if the allowance price fell below the agreed global tax for more than a certain period of time, trading partners should be allowed to levy countervailing duties on imports from countries with a low permit trading price. Third, countries could not provide tax rebates on their exports, and cap-and-trade systems would have to auction all of their allowances.

The tax should cover all the significant GHGs, insofar as is practical. The initial scheme need not cover all countries, but it should cover the countries that account for the vast majority of world emissions. All but the poorest nations should have sufficient administrative capacity to administer the tax at upstream points in the energy supply chain – that is, on the carbon content of fossil fuels.[12] The level of the tax would be set by international agreement and could be subject to periodic review every five or ten years.[13]

A carbon tax treaty would need to include monitoring and enforcement measures. The International Monetary Fund could assess whether signatory nations have passed required legislation and set up the appropriate administrative machinery to implement the tax (Agarwala). If a country were significantly and persistently out of compliance, its exports could be subject to countervailing duties in importing countries. Non-signatory countries could also be subject to countervailing duties. This possibility would provide a potent incentive for most countries to comply with the agreement, whether or not they were formal signatories.[14]

Cost-effective implementation at a global level would require the tax to be set at the same level in all countries. The abatement costs incurred by key developing countries would likely exceed by a considerable margin the maximum burden they would be willing to accept under an international agreement, at least in the near term. This could be addressed through transfers (side payments) from industrialized countries to developing countries, thereby enhancing both cost-effectiveness and distributional equity. These transfers would be from one government to another, raising concerns about possible corruption, as well as political acceptability in the industrialized world. Alternatively, distributional equity could be achieved by pairing the carbon tax agreement with a deal on trade or development that benefits these emerging economies.

[12] For example, the carbon content of oil should be taxed at refineries, natural gas should be taxed at major pipeline collection points, and coal should be taxed at mine heads or rail or barge collection points.

[13] For a thorough economic assessment of the implications of a system of harmonized domestic carbon taxes, see Bosetti, *et al.*

[14] In the section on key design issues, below, we discuss the relationship of climate policy architectures with international trade law and practices.

Coordinated national policies: linkage of national and regional tradable permit systems[15]

A new international policy architecture may be evolving on its own, based on the reality that tradable permit systems, such as cap-and-trade systems, are emerging worldwide as the favored national and regional approach. Prominent examples include the European Union's Emission Trading Scheme (EU ETS); the Regional Greenhouse Gas Initiative in the northeastern United States; and systems in Norway, Switzerland, and other nations; plus the existing global emission-reduction-credit system, the CDM. Moreover, cap-and-trade systems now appear likely to emerge as the chosen approach to reducing greenhouse gas emissions in an additional set of industrialized countries, including Australia, Canada, Japan, New Zealand, and the United States.

The proliferation of cap-and-trade systems and emission-reduction-credit systems around the world has generated increased attention and increased pressure – both from governments and from the business community – to link these systems. By linkage, we refer to direct or indirect connections between and among tradable permit systems through the unilateral or bilateral recognition of allowances or permits.[16]

Linkage produces cost savings in the same way that a cap-and-trade system reduces costs compared to a system that separately regulates individual emission sources – that is, it substantially broadens the pool of lower-cost compliance options available to regulated entities. In addition, linking tradable permit systems at the country level reduces overall transaction costs, reduces market power (which can be a problem in such systems), and reduces overall price volatility.

There are also some legitimate concerns about linkage. Most important is the automatic propagation of program elements that are designed to contain costs, such as banking, borrowing, and safety valve mechanisms. If a cap-and-trade system with a safety valve is directly linked to another system that does not have a safety valve, the result will be that both systems now share the safety valve. Given that the European Union has opposed a safety valve in its emission trading scheme, and given that a safety valve

[15] This proposed architecture was developed by Jaffe and Stavins; and supplemented by Ellerman on the European approach as a potential global model; Keohane and Raustiala on buyer liability; Hall, *et al.*; and Victor on the importance of domestic institutions; and by economic modeling from Bosetti, *et al.*

[16] As Ellerman explains, to some degree the EU ETS can serve as a prototype for linked national systems.

could be included in a future US emission trading system, this concern about the automatic propagation of cost-containment design elements is a serious one.

More broadly, linkage will reduce individual nations' control over allowance prices, emission impacts, and other consequences of their systems. This loss of control over domestic prices and other effects of a cap-and-trade policy is simply a special case of the general proposition that nations, by engaging in international trade through an open economy, lose some degree of control over domestic prices, but do so voluntarily because of the large economic gains from trade.

Importantly, there are ways to gain the benefits of linkage without the downside of having to harmonize systems in advance. If two cap-and-trade systems both link with the same emission-reduction-credit system, such as the CDM, then the two cap-and-trade systems are indirectly linked with one another. All of the benefits of linkage occur: the cost-effectiveness of both cap-and-trade systems is improved and both gain from more liquid markets that reduce transaction costs, market power, and price volatility. At the same time, the automatic propagation of key design elements from one cap-and-trade system to another is much weaker when the systems are only indirectly linked through an emission-reduction-credit system.

Such indirect linkage through the CDM is already occurring, because virtually all cap-and-trade systems that are in place, as well those that are planned or contemplated, allow for CDM offsets to be used (at least to some degree) to meet domestic obligations. Thus, indirectly linked, country- or region-based cap-and-trade systems may already be evolving into the *de facto*, if not the *de jure*, post-Kyoto international climate policy architecture.

Of course, reliance on CDM offsets also gives rise to concerns, especially as regards the environmental integrity of some of those offsets.[17] Some have recommended that a system of buyer liability (rather than seller or hybrid liability) would endogenously generate market arrangements – such as reliable ratings agencies and variations in the price of offsets according to perceived risks – that would help to address these concerns, as well as broader issues of compliance (Keohane and Raustiala). These features would in turn create incentives for compliance without resorting to ineffective inter-state punishments. In addition, a system of buyer liability gives sellers strong incentives to maintain permit quality so as to maximize the monetary value of these tradable assets.

---

[17] See section on key design issues below for an examination of ways to reform the CDM.

While in the near term, linkage may continue to grow in importance as a core element of a bottom-up, *de facto* international policy architecture, in the longer term, linkage could play several roles. A set of linkages, combined with unilateral emissions reduction commitments by many nations, could function as a stand-alone climate architecture. Such a system would be cost-effective, but might lack the coordinating mechanisms necessary to achieve meaningful long-term environmental results. Another possibility is that a collection of bottom-up links may eventually evolve into a comprehensive, top-down agreement. In this scenario, linkages would provide short-term cost savings while serving as a natural starting point for negotiations leading to a top-down agreement.[18] The top-down agreement might continue use of linked cap-and-trade programs to reduce abatement costs and improve market liquidity.

A post-2012 international climate agreement could include several elements that would facilitate future linkages among cap-and-trade and emission-reduction-credit systems. For example, it could establish an agreed trajectory of emissions caps (Frankel, "Formulas") or allowance prices, specify harmonized cost-containment measures, and establish a process for making future adjustments to key design elements. It could also create an international clearinghouse for transaction records and allowance auctions, provide for the ongoing operation of the CDM, and build capacity in developing countries. If the aim is to facilitate linkage, a future agreement should also avoid imposing "supplementarity" restrictions that require countries to achieve some specified percentage of emission reductions domestically.

Key design issues in international policy architecture

Regardless which overall international policy architecture is ultimately chosen, a number of key design issues will stand out as particularly important. Based on research carried out under the auspices of the Harvard Project on International Climate Agreements, we identify key lessons for five issues and elements relevant to a post-2012 international agreement: burden-sharing, technology transfer, CDM reform, addressing deforestation, and

---

[18] Carraro (2007) and Victor (2007) also describe the potential for trading to emerge organically as a result of linking a small set of domestic trading programs. This evolution would be analogous to the experience in international trade in goods and services, in which a small number of countries initially reached agreement on trade rules governing a small set of goods. As trust built on these initial experiences, trading expanded to cover more countries and more goods, a process that eventually provided the foundation for a top-down authority in the form of the World Trade Organization.

making global climate policy compatible with global trade policy. All five of these issues are relevant to the relationship between global climate policy and economic development (Wirth).

Burden-sharing in an international climate agreement

The most challenging aspect of establishing a post-Kyoto international climate regime will be reaching agreement on burden-sharing among nations that will be explicitly or implicitly part of the adopted regime. In this context, the interface between global climate policy and economic development becomes particularly important.

One approach to thinking about this issue is to start by focusing on what is politically possible and to identify an allocation of responsibility – with appropriate changes over time – that makes every country feel that it is doing only its fair share (Frankel, "Formulas"). A common thread in many discussions about "fair," long-term burden-sharing is the desirability of gradually moving all countries toward equal per capita emissions.[19] As a long-term outcome this would be consistent with what many people, from diverse perspectives, regard as ultimately equitable (Agarwala; Cao; and Frankel, "Formulas"), although others have noted that if the goal is greater equity in the distribution of wealth, directly targeting wealth redistribution would be more effective (Posner and Sunstein).

More broadly, the three-element formula proposed by Frankel for setting evolving country-level emissions targets has the virtue of recognizing the industrialized countries' historic responsibility for GHG emissions (Agarwala and Somanathan) and does not reward countries for previous lack of action. Furthermore, this time-path of evolving commitments reflects the reality that, in the short term, developing countries value their economic growth more than future environmental conditions (Victor). But by providing for increased participation by developing countries over time, this approach also recognizes that it will be impossible to stabilize atmospheric GHG concentrations unless rapidly growing developing countries take on an increasingly meaningful role in reducing global emissions (Blanford, *et al.*; Bosetti, *et al.*; Clarke, *et al.*; Cooper; Hall, *et al.*; and Jacoby, *et al.*). The real test lies in whether domestic constituencies in the developed world will perceive such agreements as fair.

---

[19] Somanathan argues that although an effective solution to climate change will require the cooperation of developing countries, achieving near-term GHG reductions in these countries will be neither feasible nor desirable because of their other priorities for economic and social development.

Technology transfer in an international climate agreement[20]

Achieving long-term climate change policy goals will require a remarkable ramp-up in the innovation and deployment of energy-efficient and low-carbon technologies in an environment that is already experiencing substantial increases in investment (Aldy and Stavins 2008c; Newell; Clarke, *et al.*).[21] Transitioning away from fossil fuels as the foundation of industrialized economies and as the basis for development in emerging economies will likely necessitate a suite of policies to provide the proper incentives for technological change (Somanathan). Two principal categories of policies are potentially important to drive the invention, innovation, commercialization, diffusion, and utilization of climate-friendly technologies: (1) international carbon markets and other pricing strategies and (2) non-price mechanisms, including various means of technology transfer to developing countries and coordinated innovation and commercialization programs.

International carbon markets and technology transfer

The most powerful tool for accelerating the development and deployment of climate-friendly technologies will be policies that affect the current and expected future prices of fossil fuels relative to lower-carbon alternatives. By setting a price on GHG emissions and thereby raising the price of conventional fossil fuels and energy-intensive production practices, these policies – which are at the core of several proposed international climate policy architectures – will induce investment in less emissions-intensive technologies. Cap-and-trade programs; emission-reduction-credit systems, such as the CDM; and harmonized domestic carbon taxes can thus create incentives for emission mitigation projects in industrialized and developing countries alike.

Given the long lifetimes of many emissions-intensive capital assets – power plants may operate 50 years or more, building shells may last 100 years – long-term carbon price signals may be necessary to allow the owners of such capital to form appropriate expectations and alter the nature of their investments. Blunt policy instruments such as performance standards or

---

[20] Below we address technology transfer in the context of efforts to reform the CDM.

[21] The International Energy Agency forecasts more than $20 trillion of investment in the global energy infrastructure between now and 2030. Some of this accelerated investment is evident in China, where one out of six coal-fired power plants is less than three years old. But the investment is not universal – populations in least developed countries still suffer from lack of access to power and basic energy poverty that can inhibit advances along a variety of development measures (Aldy and Stavins 2008c).

bans on carbon-intensive products can also induce innovation, but such approaches are typically less efficient.

Pricing carbon can leverage foreign direct investment to promote less carbon-intensive development. For example, some CDM projects have resulted in the deployment of renewable power, such as wind farms, as an alternative to coal-fired power generation. Other CDM projects have been criticized for rewarding minor process modifications that do not involve substantial investment in new technologies, such as the manufacture of fluorinated refrigerants. Some countries may also consider CDM participation a substitute for taking further mitigation actions or even use the CDM to justify weakening policies in other areas. More broadly, reforming the CDM could facilitate more substantial transfers of technology (Keeler and Thompson; Hall, *et al.*; and Teng, *et al.*). We consider such approaches below.[22]

In any event, putting a price on carbon may not facilitate new investment flows and associated technology transfers to developing countries with weak market institutions. If a country has difficulty attracting capital generally, changing the relative prices of carbon-intensive and carbon-lean capital will not resolve this problem. In this case, additional policy interventions would be required to stimulate the transfer of technology to developing countries. Also, while putting a price on carbon will draw more resources into low-carbon technology R&D, it will not be sufficient to fully overcome the general disincentive for private-sector investments in R&D. This is because undertaking R&D effectively produces new knowledge, and this knowledge is a public good. Once the knowledge exists, it is difficult for firms to prevent others from sharing its benefits (although patent law provides some protection). Since innovating firms cannot capture all the benefits of their R&D efforts, they tend to under-invest in such activities. Thus, additional policies are needed to promote the public- and private-sector innovation that will be required to ensure that a next generation of climate-friendly technologies is available for deployment.

Additional technology policy in post-2012 international climate agreements

The next international climate agreement can provide several mechanisms to facilitate the development and deployment of climate-friendly technologies (Aldy and Stavins 2008c; Newell; and Somanathan; Clarke, *et al.*). First, the agreement can provide a venue for countries to pledge resources for technology transfer and for R&D activities (Newell). The agreement could

---

[22] An alternative to reforming the CDM that could also facilitate greater technology transfer is to establish climate accession deals with individual developing countries (Victor).

also codify such pledges as commitments, on par with commitments to limit emissions (as in Annex B of the Kyoto Protocol). Besides negotiating a given level of financial commitment, developed and emerging countries could explicitly articulate how they mean to meet their commitments, thereby promoting credibility and trust in the agreement. This could take the form of identifying a specific revenue stream (for example, auction revenues from a cap-and-trade program) that would be adequate and reliable for supporting financial pledges.

Financing technology transfer will require coordination and agreement on principles for allocating resources. An institutional home for clean technology funds may be necessary, in which case the international policy community will need to decide whether to centralize such efforts in a new institution, rely on an existing international institution, or manage the program through a decentralized array of national institutions. Likewise, some agreement on the means for coordinating R&D activities will have to be considered in identifying the appropriate institutional design.

A framework for coordinating and augmenting climate technology R&D could be organized through a UNFCCC Expert Group on Technology Development, supported by the International Energy Agency (Newell). Broadening IEA participation to include large non-OECD energy consumers and producers could also facilitate such coordination. An agreement could include a process for reviewing country submissions on technology development and for identifying redundancies, gaps, and opportunities for closer collaboration. A fund for cost-shared R&D tasks and international technology prizes could be established to provide financing for science and innovation objectives that are best pursued in a joint fashion. The agreement could also include explicit targets for increased domestic R&D spending on GHG mitigation.

An independent mechanism for reviewing policies that affect technological development and deployment may benefit these efforts. Rigorous, third-party review of all nations' policies and financing mechanisms could support coordinated, international efforts by providing an authoritative assessment of the comparability of effort among participating countries. This could include reviews of financial contributions by large countries, analyses of the effectiveness of technology transfer activities, and identification of the best policy practices being implemented around the world. Such reviews could be undertaken by an existing international institution or may require the creation of a new, professional bureaucracy focused on this single surveillance task. The same institution or mechanisms could also help to assess the comparability of efforts on mitigation, adaptation, and other elements of an international agreement.

In addition to strengthening incentives, barriers to climate-friendly technology transfer could be lowered through a World Trade Organization (WTO) agreement to reduce tariff and non-tariff barriers to trade in environmental goods and services (Newell). Development and harmonization of technical standards – which could be undertaken by international standards organizations in consultation with the IEA and WTO – could further reduce impediments to technology transfer and accelerate the development and adoption of climate-friendly innovations.

Finally, an international climate policy architecture could provide positive incentives for developing and emerging economies to pursue good policy practices. For example, conditioning access to climate technology funds on the implementation of domestic "no regrets" climate policies could substantially increase the "climate return" to technology fund resources. Alternatively, access to clean technology funds could be scaled based on the extent of policy action in developing and emerging economies – as governments implement more climate-friendly policies they could access a larger pool of resources. Such determinations could be made on the basis of independent, expert reviews of countries' climate and energy policies.

Reforming the Clean Development Mechanism

One of the important principles identified by our research teams is that, because there are very large start-up costs for creating new institutions, consideration should be given to maintaining existing institutions, such as the CDM, and improving them rather than abandoning them (Hall, *et al.*; Karp and Zhao; Keeler and Thompson; and Teng, *et al.*).

As we emphasized earlier, serious criticisms have been leveled at the CDM in its current form: because the CDM is an emission-reduction-credit system (not a cap-and-trade system), the concern is that it may credit emission reductions that are not truly additional. There have been numerous calls to address the CDM's problems by putting in place criteria and procedures to increase the likelihood that certified offset credits represent emission reductions that are truly "additional, real, verifiable, and permanent." While such reforms would have merit if they were effective, there are a number of alternative, more dramatic changes in the CDM that merit consideration.[23]

---

[23] We noted above the possibility of addressing the problems of the CDM through a system of buyer (rather than seller or hybrid) liability, in order to generate market arrangements that would help address these critiques, such as reliable ratings agencies, and variations in the price of offsets according to perceived risks (Keohane and Raustiala). This approach would give sellers incentives to maintain permit quality to maximize the monetary value of these tradable assets.

Improved, expanded, and focused GHG offsets

One promising approach would involve less emphasis on strict ton-for-ton accounting and more emphasis on a range of activities that could produce significant long-term benefits (Keeler and Thompson). There are five key elements of this proposal. First, the criteria for CDM offsets would be changed from "real, verifiable, and permanent reductions" to "actions that create real progress in developing countries toward mitigation and adaptation." The reasoning behind this change is that strict, project-based accounting rules, while intended to protect the environmental integrity of trading programs, have increased transaction costs and thereby limited the utility of the CDM. The argument is that developing country actions are more important than the sanctity of short-term targets in making real progress on mitigating climate change risk.

Second, this proposal would make a significant share of industrialized country commitments (whether international or domestic) achievable through offset payments to developing countries. If industrialized countries aimed to purchase offset credits equivalent to at least 10 percent of their overall emissions targets, they would greatly expand the flow of resources available to support developing country actions. Third, a specified portion of offset credits (perhaps 50 percent) would be sold up front, and the proceeds would be put in a fund for supporting investments in projects throughout the developing world. By allowing greater flexibility to support large-scale or non-standard projects, this approach could increase the geographic diversity of mitigation activities and reduce transaction costs.

Fourth, international negotiations would be focused on developing guidelines for an international offsets program. Key issues to be addressed would include criteria for eligible activities, policies, and investments; requirements for documentation or accountability; mechanisms for ex post adjustment; criteria for the distribution of funds; and set-asides, if any, for particular types of projects or technologies. Fifth, clearly delineated tasks would be delegated to new and existing institutions for the purpose of managing and safeguarding the offsets program.

Such reform of the CDM could facilitate more substantial transfers of technology (Aldy and Stavins 2008c). In addition, creating a list of pre-approved technologies could lower the transaction costs of the review and certification process, and thus encourage more projects. Expanding the coverage of the CDM from specific projects to an entire industry, such as the power sector, could promote the exploitation of all low-cost mitigation opportunities in that country's industry, some of which may be too small to be proposed on a project-by-project basis. Modifying the CDM to include policies, as well as projects, could also stimulate further investment in low-carbon technologies. For example, credits could be awarded for implementing vehicle fuel-economy standards, reducing fossil-fuel subsidies, or enforcing land-tenure rules that slow deforestation.

On the other hand, efforts to improve the performance of the CDM as a means for transferring climate-friendly technologies to developing countries also confront some major challenges (Aldy and Stavins 2008c). First, the difficulty of demonstrating additionality in a project context may become even greater in an industry or policy context – that is, the problem of constructing a project-based counterfactual (what would have happened anyway) becomes a similar counterfactual estimation problem, only at the more complicated level of a broader industry or policy. Second, limits imposed by industrialized countries on the volume of CDM credits that can enter their carbon markets will lower credit prices and discourage some new technology investment. Third, the CDM may create disincentives for some emerging economies to take on more substantial action domestically or make commitments as part of an international agreement.

If the transfer of climate-friendly technology from developed to developing countries is necessary to address climate change, then some have argued that the objective of a revamped CDM should not be primarily to capture inexpensive mitigation opportunities ("low hanging fruit"), but rather to spur new and replicable technology transfer from developed to developing countries. Consistent with this notion, some have proposed a "Technology CDM" under which technology transfer would be the only emissions-reducing activity for which credits would be awarded (Teng, *et al.*). This would offer the opportunity to strengthen the technology transfer effects of the CDM in the near term without redesigning the whole system.

Climate accession deals

Others have taken the early limitations of the CDM as evidence that a fundamentally different approach will be needed to make real progress. One proposal that reflects this view is for climate accession deals to be employed as a new strategy for engaging developing countries (Victor). This approach builds on two premises: first, that developing nations value economic growth far more than they value future global environmental conditions and second, that many governments of developing nations lack the administrative ability to control emissions.

Under this proposal, climate accession deals would be negotiated on a country-by-country basis. An individual accession deal would essentially consist of a set of policies that are tailored to gain maximum leverage on a single developing country's emissions, while still aligning with its interests and capabilities. Industrialized countries would support each accession deal by providing specific benefits, such as financial resources, technology, administrative training, or security guarantees. Because these deals would be complex to engineer, they should be few in number and focused on nations with extremely high potential for reducing emissions.

A given developing country would bid a variety of policies and programs that make sense for its development trajectory. Its bid would include information on existing barriers (e.g., funding, technology, access to international institutions). Subsequent international negotiations would determine the resources that industrialized nations would provide to that country and the metrics for assessing compliance. Accession deals could assist developing countries in adhering to global norms for GHG abatement efforts, akin to trade accession deals that promote adherence to a consistent set of trade rules.

Compared to conventional approaches, accession deals could have several advantages (Victor). First, they would be anchored in host countries' interests and capabilities. Second, they would be limited in number and could yield a significant degree of leverage while minimizing external investment. Third, they would engage private enterprise and government ministries that are beyond the environmental and foreign affairs ministries. Fourth, such accession deals would be replicable and scalable. Where they succeed, they could offer templates for similar deals in other countries.

Addressing deforestation in an international climate agreement[24]

Forest carbon flows comprise a significant part of overall global GHG emissions, with deforestation contributing between 20 and 25 percent of net emissions. Worldwide, the amount of CO_2 sequestered in forest vegetation is approximately 1,300 billion tons, compared with annual industrial CO_2 emissions of 23 billion tons. Thus, changes to forests can have enormous impacts on the global carbon cycle, which implies in turn that forest carbon management ought to be an element of the next international agreement on climate change. A promising path forward could involve taking a "national

[24] This section draws on Plantinga and Richards.

inventory" approach, in which nations receive credits or debits for changes in forest cover relative to a measured baseline (Plantinga and Richards).

Three basic approaches could be taken to address deforestation in an international climate agreement. The first approach, currently used by the CDM, relies on project-level accounting. Under this system, individual landowners can apply for credits for net increases in carbon stored in forests on their land. Once the permitting authority verifies that the claimed sequestration is valid, the landowner can sell the credits in allowance markets. But experience has shown that such project-by-project accounting faces serious challenges, especially in establishing the counterfactual baseline against which to evaluate projects. This additionality problem is compounded by problems of leakage (the off-site effects of projects), permanence (the potential for future changes or events to result in the release of sequestered carbon), and adverse selection problems (the most profitable projects, which are most likely to occur anyway, are also the most likely to be credited under project-oriented CDM).

A second policy approach would "delink" forest carbon programs from emission allowance systems. Rather than focusing on carbon credits, the program would focus on inputs such as policies to discourage deforestation, programs to encourage the conversion of marginal agricultural land to forests, and projects to better manage forests in forest-rich countries. These commitments could be funded by overseas development aid, international institutions, or through a separate climate fund. A delinked system would have some advantages in terms of lower transaction costs and by virtue of opening separate negotiations over international forest sequestration and energy emissions. But this approach would also have two serious disadvantages. First, incentives for forest-based carbon sequestration would be diminished and participating countries might shift their attention from assuring positive carbon outcomes to attracting project funding. Second, decoupling the forest carbon program from cap-and-trade systems removes one of the best sources of funding to promote land-use changes – emitters seeking lower-cost options to reduce their net emissions.[25]

A third, more promising approach is national inventory accounting. Under this approach, nations would conduct periodic inventories of their forest carbon stock. The measured stock would be compared with a pre-negotiated baseline to determine the offset credits that can be redeemed, or debits that must be covered, in the permit market. With this approach, national

---

[25] Bosetti, *et al.* find that including credits for deforestation in a global cap-and-trade system reduces costs significantly.

governments, rather than project developers, pursue carbon sequestration activities through the implementation of domestic policies. International negotiations would determine the reference or baseline stock of stored forest carbon. These negotiations could be used to address equity issues, as well as provide incentives for countries – in particular, countries with declining stocks – to participate in the agreement.

A national inventory approach would greatly reduce the problems that plague the CDM's project-by-project approach. It could also provide comprehensive coverage of changes to forest carbon stocks and be applied equally to all participating countries and to all measurable changes in forest carbon stocks. There are also some reasonable concerns about this approach. First, the scope of carbon sequestration activities is limited to those that can be measured. Second, the approach provides incentives for governments, not private project developers, which may be a disadvantage in countries with weak institutions, high levels of corruption, or powerful special interest groups. Third, problems with additionality, permanence, etc. may resurface with – and reduce the effectiveness of – domestic carbon sequestration policies pursued by national governments.[26]

Making global climate policy compatible with global trade policy[27]

Global efforts to address climate change could be on a collision course with global efforts through the WTO to reduce barriers to trade (Frankel, "Trade"). With different countries likely to adopt different levels of commitment to climate change mitigation, the concern arises that carbon-intensive goods or production processes could shift to countries that do not regulate GHG emissions. This leakage phenomenon is viewed as problematic – by environmentalists because it would undermine emission-reduction objectives and by industry leaders and labor unions because it could make domestic products less competitive with imports from nations with weaker GHG regulations. Thus, various trade measures – including provisions for possible penalties against imports from countries viewed as non-participants – have been included in some climate policy proposals in the United States and Europe, as well as in proposals for a post-2012 international policy architecture (Jaffe and Stavins; Karp and Zhao).

[26] A delinked, input-based approach could be used as an interim strategy while the scientific community works to develop the measurement capacity necessary to support national inventories.

[27] This section draws extensively on Frankel ("Trade"); supplemented by Karp and Zhao on trade sanctions; and Newell, and Hall, *et al.* on subsidies for international transfers.

The widespread impression that the WTO is hostile to environmental concerns has little basis in fact. The WTO's founding Articles cite environmental protection as an objective; environmental concerns are also explicitly recognized in several WTO agreements. Recent WTO rulings support the principle that countries not only have the right to ban or tax harmful products, but – perhaps more critically – that trade measures can be used to target processes and production methods, provided they do not discriminate between domestic and foreign producers. The question is how to address concerns about leakage and competitiveness in a way that does not run afoul of WTO rules and avoids derailing progress toward free trade and climate goals alike.

Future national-level policies to address climate change may be expected to include provisions that target carbon-intensive products from countries deemed to be making inadequate efforts. These provisions need not violate sensible trade principles and WTO rules, but there is a danger that in practice they will. The kinds of provisions that would be more likely to conflict with WTO rules and provide cover for protectionism include the following: (1) unilateral measures applied by countries that are not participating in the Kyoto Protocol or its successors; (2) judgments made by politicians vulnerable to political pressure from interest groups for special protection; (3) unilateral measures that seek to sanction an entire country, rather than targeting narrowly defined energy-intensive sectors; (4) import barriers against products that are further removed from carbon-intensive activity, such as firms that use inputs that are produced in an energy-intensive process; and (5) subsidies – whether in the form of money or extra permit allocations – to domestic sectors that are considered to have been put at a competitive disadvantage.

By contrast, border measures that are more likely to be WTO-compatible include either tariffs or (equivalently) requirements for importers to surrender tradable permits designed with attention to the following guidelines: (1) trade measures follow some multilaterally-agreed set of guidelines among countries participating in the emission targets of the Kyoto Protocol and/or its successors; (2) judgments about which countries are complying or not, which industries are involved and their carbon content, and which countries are entitled to respond with border measures are made by independent panels of experts; (3) measures are applied only by countries that are reducing their emissions against countries that are not doing so – either as a result of their refusal to join an agreement or their failure to comply; and (4) import penalties target fossil fuels, and five or six of the most energy-intensive major industries that produce manufactured bulk goods: aluminum, cement, steel, paper, glass, and perhaps iron and chemicals.

The economics and the laws governing the interaction of trade and environmental policy are complex, and a multilateral regime is needed to guide the development of trade measures intended to address concerns about leakage and competitiveness in a world where nations have different levels of commitment to GHG mitigation. Ideally, this regime would be negotiated along with a successor to the Kyoto Protocol that sets emission-reduction targets for future periods. If that process takes too long, however, it might be useful in the shorter term for a limited set of countries to enter into negotiations to harmonize guidelines for border penalties, ideally in informal association with the secretariats of the UNFCCC and the WTO.

Conclusion

Great challenges confront the community of nations seeking to establish an effective and meaningful international climate regime for the post-2012 period. But some key principles, promising policy architectures, and guidelines for essential design elements have begun to emerge.

Climate change is a global commons problem, and therefore a cooperative approach involving many nations will be necessary to address it successfully. Since sovereign nations cannot be compelled to act against their wishes, successful treaties must create adequate internal incentives for compliance, along with external incentives for participation. A credible global climate change agreement must be: (1) equitable; (2) cost-effective; (3) able to facilitate significant technological change and technology transfer; (4) consistent with the international trade regime; (5) practical, in the sense that it builds – where possible – on existing institutions and practices; (6) attentive to short-term achievements, as well as medium-term consequences and long-term goals; and (7) realistic. Because no single approach guarantees a sure path to ultimate success, the best strategy may be to pursue a variety of approaches simultaneously.

The Harvard Project on International Climate Agreements does not endorse a single international climate policy architecture. Rather, we have highlighted four potential frameworks for a post-Kyoto agreement, each of which is promising in some regards and raises important issues for consideration. One calls for emissions caps established using a set of formulas that assign quantitative emissions limits to countries through 2100. These caps would be implemented through a global system of linked national and regional cap-and-trade programs that would allow for trading among firms and sources. A second potential framework would instead rely on a system of linked international agreements that separately address mitigation in various sectors and gases, along with issues like adaptation, technology research and development, and geoengineering. A third architecture would consist of harmonized domestic taxes on emissions of GHGs from all sources, where the tax or charge would be internationally adjusted from time to time, and each country would collect and keep the revenues it generates. Fourth, we discussed an architecture that – at least in the short term – links national and regional tradable permit systems only indirectly, through the global CDM. We highlight this option less as a recommendation and more by way of recognizing the structure that may already be evolving as the *de facto* post-Kyoto international climate policy architecture.

Regardless of which overall international policy architecture is chosen, a number of key design issues will stand out as particularly important, including burden-sharing, technology transfer, CDM reform, addressing deforestation, and making global climate policy compatible with global trade policy. All of these issues involve the relationship between global climate policy and economic development, and all are under careful investigation as part of the Harvard Project.

As the Harvard Project on International Climate Agreements moves forward, we continue to draw upon leading thinkers from academia, private industry, government, and non-governmental organizations around the world. We also continue to work with our research teams around the world, and meet in a wide variety of venues with those who can share their expertise and insights. We look forward to receiving input regarding all elements of our work – including feedback on the proposals and analyses which constitute the content of this *Summary for Policymakers*.

References

Aldy, Joseph E., Scott Barrett, and Robert N. Stavins (2003). "Thirteen Plus One: A Comparison of Global Climate Policy Architectures," *Climate Policy* 3(4): 373-397.

Aldy, Joseph E. and Robert N. Stavins, eds. (2007). *Architectures for Agreement: Addressing Global Climate Change in the Post-Kyoto World.* New York: Cambridge University Press.

Aldy, Joseph E. and Robert N. Stavins (2008a). "Climate Policy Architectures for the Post-Kyoto World," *Environment* 50(3): 6-17.

Aldy, Joseph E. and Robert N. Stavins (2008b). "Economic Incentives in a New Climate Agreement," Prepared for The Climate Dialogue, Hosted by the Prime Minister of Denmark, May 7–8, 2008, Copenhagen, Denmark. Cambridge, Mass.: Harvard Project on International Climate Agreements, May 7.

Aldy, Joseph E. and Robert N. Stavins (2008c). "The Role of Technology Policies in an International Climate Agreement." Prepared for The Climate Dialogue, Hosted by the Prime Minister of Denmark, September 2–3, 2008, Copenhagen, Denmark. Cambridge, Mass.: Harvard Project on International Climate Agreements, September 3.

Aldy, Joseph E. and Robert N. Stavins, eds. (2009). *Post-Kyoto International Climate Policy: Implementing Architectures for Agreement.* New York: Cambridge University Press.

Carraro, Carlo (2007). "Incentives and Institutions: A Bottom-Up Approach to Climate Policy," in Aldy and Stavins (eds.), pp. 161-172.

Frankel, Jeffrey (2007). "Formulas for Quantitative Emission Targets," in Aldy and Stavins (eds.), pp. 31-56.

Hahn, Robert W. and Robert N. Stavins (1999). *What Has the Kyoto Protocol Wrought? The Real Architecture of International Tradable Permit Markets.* Washington, DC: American Enterprise Institute Press.

Information Office of the State Council (2008). "China's Policies and Actions for Addressing Climate Change," white paper published by the government of the People's Republic of China, October 29. Available at http://china.org.cn/government/news/2008-10/29/content_16681689.htm.

Jiang, Kejun (2008). "Opportunities for Developing Country Participation in an International Climate Change Policy Regime," Discussion Paper 08-26. Cambridge, Mass.: Harvard Project on International Climate Agreements, November.

McKibbin, Warwick J. and Peter J. Wilcoxen (2007). "A Credible Foundation for Long-Term International Cooperation on Climate Change," in Aldy and Stavins (eds.), pp. 185-208.

Michaelowa, Axel (2007). "Graduation and Deepening," in Aldy and Stavins (eds.), pp. 81-104.

Pan, Yue (2007). *Thoughts on Environmental Issues.* Beijing: China Environmental Culture Promotion Association.

Pizer, William A. (2007). "Practical Global Climate Policy," in Aldy and Stavins (eds.), pp. 280-314.

Schmalensee, Richard (1998). "Greenhouse Policy Architectures and Institutions," in W. D. Nordhaus (ed.), *Economics and Policy Issues in Climate Change*, Washington, DC: Resources for the Future Press, pp. 137-158.

Victor, David G. (2007). "Fragmented Carbon Markets and Reluctant Nations: Implications for the Design of Effective Architectures," in Aldy and Stavins (eds.), pp. 133-160.

Appendix 1 Summaries of Research Initiatives Harvard Project on International Climate Agreements

Section I

Alternative International Policy Architectures

An elaborated proposal for global climate policy architecture: specific formulas and emission targets for all countries in all decades

Jeffrey Frankel

Overview

This proposal builds on the foundations of the Kyoto Protocol, but strengthens it in important ways. It attempts to solve the most serious deficiencies of Kyoto: the absence of long-term targets, the absence of participation by the United States and developing countries, and the lack of motivation for countries to abide by their commitments. Although there are many ideas to succeed Kyoto, virtually all the existing proposals are based either on science (e.g., capping global concentrations at 450 ppm) or on economics (weighing the economic costs of aggressive short-term cuts against the long-term environmental benefits). The plan for emissions reductions proposed in this paper is more practical because it is partly based on politics, in addition to science and economics.

Discussion

The proposal calls for an international agreement to establish a global cap-and-trade system. The emissions caps are set using formulas that assign quantitative emissions limits to countries in every year until 2100. Three political constraints are particularly important in developing the formulas. First, developing countries are not asked to bear any cost in the early years. Second, even later, developing countries are not asked to make any sacrifice that is different from the earlier sacrifices of industrialized countries, accounting for differences in incomes. Third, no country is asked to accept targets that cost it more than 5 percent of GDP in any given year.

Under the formulas, rich nations begin immediately to make emissions cuts. Developing countries agree to maintain their business-as-usual emissions in the first decades, but over the longer term agree to binding targets that ultimately reduce emissions below business as usual. This structure precludes energy-intensive industries from moving operations to developing countries (so-called "carbon leakage") and gives industries a more even playing field. However, it still preserves developing countries' ability to grow their economies, and they can raise revenue by selling emission permits. In later decades, once developing countries cross certain income and emissions thresholds, their emissions targets become stricter, following a numerical formula. However, these emissions cuts are no greater than the cuts made by rich nations earlier in the century, accounting for differences in per capita income, per capita emissions, and baseline economic growth.

This system of targets results in a world price of carbon dioxide that reaches $30 per ton in 2020, $100 per ton in 2050, and $700 per ton in 2100, according to economic simulations using the WITCH climate model. Most countries sustain economic losses that are under 1 percent of GDP in the first half of the century, but then rise toward the end of the century. Atmospheric concentrations of CO_2 stabilize at 500 ppm in the last quarter of the century, and world temperatures increase by about 3 degrees Celsius.

Key findings and recommendations

- *Any future climate agreement must comply with six important political constraints.* First, the United States will not commit to quantitative targets if China and other major developing countries do not commit to quantitative targets at the same time, due to concerns about economic competitiveness and carbon leakage. Second, China and other developing countries will not make sacrifices different in character from those made by richer countries that have gone before them. Third, in the long run, no country can be rewarded for having "ramped up" its emissions high above the levels of 1990. Fourth, no country will agree to participate if, in any year, the present discounted value of its future expected costs is more than 1 percent of GDP. Fifth, no country will abide by targets that cost it more than 5 percent of GDP in any year. Sixth, if one major country drops out, others will become discouraged and the system may unravel.
- *Future emissions caps should be determined by a formula that incorporates three elements: a Progressivity Factor, a Latecomer Catch-up Factor, and*

a Gradual Equalization Factor. The Progressivity Factor requires richer countries to make more severe cuts relative to their business-as-usual emissions. The Latecomer Catch-up Factor requires nations that did not agree to binding targets under Kyoto to make gradual emissions cuts to account for their additional emissions since 1990. This factor prevents latecomers from being rewarded with higher targets or from being given incentives to ramp up their emissions before signing the agreement. Finally, the Gradual Equalization Factor addresses the fact that rich countries are responsible for most of the carbon dioxide currently in the atmosphere. During each decade of the second half of the century, this factor moves per capita emissions in each country a small step in the direction of the global average of per capita emissions.

Conclusion

The framework here allocates emission targets across countries in such a way that every country is given reason to feel that it is only doing its fair share. Furthermore, the framework – a decade-by-decade sequence of emission targets determined by a few principles and formulas – is flexible enough that it can accommodate major changes in circumstances during the course of the century.

Author affiliation

Jeffrey Frankel is the James W. Harpel Professor of Capital Formation and Growth at the Harvard Kennedy School.

Appendix 1.2   The EU Emission Trading Scheme: a prototype global system?

A. Denny Ellerman

Overview

As the world's first multinational cap-and-trade system for regulating green-house gas (GHG) emissions, the European Union Emission Trading Scheme (EU ETS) can be seen as a prototype for an eventual global climate regime. This paper draws on the first four years of experience with the EU ETS to develop insights about the challenges that can be expected to emerge in a broader program and to suggest potential solutions.

Discussion

The ETS was launched in 2005 to help EU member states meet their Kyoto Protocol commitments. It covers CO_2 emissions from power plants and other large industrial facilities, which together account for approximately 40 percent of the EU's total GHG inventory. Following an initial three-year trial phase, the program entered the first of the succeeding 'real' trading periods in 2008.

Despite numerous difficulties, the EU has largely succeeded in establishing a functioning cap-and-trade system that is generating concrete price signals for reducing CO_2 emissions. Strikingly, the number of nations participating in the ETS has doubled from fifteen, when the program was conceived, to thirty today. The current program includes countries that vary widely in their level of economic development, institutional capacity, regulatory history, and domestic commitment to climate-change mitigation. This diversity and the sovereign status of EU member states makes the EU ETS better suited to serve as a model for a global system than might first appear.

Key findings and recommendations

- *An initial trial period was invaluable in providing the opportunity to correct widespread data deficiencies and develop needed institutional capacities.* As a result, it was possible to launch the first 'real' trading period of the ETS with national emission budgets based on verified facility-level emissions data, functioning monitoring protocols and registries, and full-fledged participation by all members. Because allowances issued during the trial period could not be banked or borrowed, early problems (such as the over-allocation of allowances) were prevented from spilling over into later trading periods.

- *The centralized functions provided by the European Commission will be equally essential to the success of a global system, but it is not yet clear what organization could step in to fill that role.* Although individual states are responsible for emissions monitoring, reporting, verification, and enforcement under the ETS, the European Commission plays a critical role in approving national budgets, establishing common registry protocols, and providing information and technical assistance. What entity could fill these needs for a global system remains a crucial and as yet unanswered question.

- *The broader benefits of EU membership help account for the willingness of less committed nations to join the ETS. Such 'club' benefits may be important for securing broad-based participation in a global system.* While a number of newer EU members are unhappy with the ETS and several have mounted formal legal challenges to proposed emissions budgets, appeals are being pursued through common European institutions, and no nation has yet withdrawn from the program. Apparently, the benefits of EU membership continue to outweigh the disadvantages of participation. Similarly effective inducements for opting in (and staying in) will need to exist in a global system.

- *Reconciling increased stringency, differentiation, and harmonization pose a major challenge for any multinational GHG trading system.* The evolution of the EU ETS suggests that increasing stringency requires greater differentiation of responsibilities among countries of differing circumstances at the same time that participants are calling for greater harmonization in allowance allocations, a goal that is at odds with differentiation so long as allowances are freely allocated. These conflicting goals are to be reconciled through full auctioning, with differentiated allocation of auction rights to member states.

- *The public may be more willing to accept international trade in emissions allowances than previously imagined.* There has long been a concern that large outflows of capital to purchase allowances from other countries could prove politically problematic. So far this has not been an issue in the EU ETS, perhaps because international transfers of allowances are (a) relatively small and (b) dwarfed by trade in other goods and services.

Conclusion

Interestingly, the problems that are often seen as most intractable for a global trading system – institutional readiness and public acceptance – haven't yet appeared in Europe. Rather, the greater challenges may lie in developing an effective centralized authority, devising side benefits to encourage participation, and dealing with the interrelated issues of harmonization, differentiation, and stringency. The EU ETS is not perfect, nor does it provide a perfect prototype for a global system, which would surely diverge in important respects from the European model. Nevertheless, the EU example is likely to continue to be highly instructive as policymakers consider the larger and more difficult task of constructing a global trading system.

Author Affiliation

A. Denny Ellerman is Senior Lecturer at the Massachusetts Institute of Technology's Sloan School of Management.

Appendix 1.3 Linkage of tradable permit systems in international climate policy architecture

Judson Jaffe and Robert N. Stavins

Overview

Tradable permit systems have emerged around the world as a preferred instrument for reducing emissions of greenhouse gases. Because linking tradable permit systems can reduce compliance costs, there is great interest in doing so. This paper examines the benefits and concerns associated with linkage and analyzes the near-term and long-term roles that linkage may play in a future international climate policy architecture.

Discussion

There are two types of tradable permits systems: cap-and-trade systems, in which a government issues allowances that firms must obtain to emit greenhouse gases, and emission-reduction-credit systems, in which firms can earn credits by voluntarily reducing emissions. The opportunity to trade allowances or credits within a tradable permit system introduces flexibility and economic incentives that can minimize emission reduction costs. However, absent linkages between systems, some emission reductions required in one system may be more costly than reduction opportunities that remain untapped in another system, leaving cost-saving opportunities unrealized.

Due to the increasingly likely prospect of a world with multiple greenhouse gas tradable permit systems, attention has turned to the questions of whether and how to link these systems. Direct linkages occur when a system's regulatory authority allows firms to use allowances or credits from another system for compliance purposes in its own system. In turn, direct linkages can lead to indirect linkages. For example, cap-and-trade systems can become indirectly linked with one another if each establishes a direct link with a common emission-reduction-credit system, such as the Clean Development Mechanism (CDM).

Key findings and recommendations

- *Linkages can significantly reduce the cost of achieving global emission targets and can offer other important benefits.* Allowance or credit trading across systems can generate cost savings by allowing higher-cost reductions in one system to be replaced by lower-cost reductions in another system. Linkages can also reduce allowance price volatility by improving market liquidity and can allow for "common but differentiated responsibilities" across systems without increasing the cost of achieving global emission targets.

- *At the same time, some linkages can raise legitimate concerns.* For example, direct linkages with other cap-and-trade systems can reduce a country's control over allowance prices in its own system and can lead to automatic propagation of cost containment measures – banking, borrowing, and safety-valves – from one system to another. Also, linkages with emission-reduction-credit systems may reduce the environmental effectiveness of a cap-and-trade system if the credit system credits some emission reductions that are not truly additional.

- *In the near term, indirect linkages among cap-and-trade systems through the CDM or some other global emission-reduction-credit system may be most promising.* Direct linkages between cap-and-trade systems may require harmonization of key system design elements because of the automatic propagation of cost-containment measures and other consequences of such linkages. By contrast, indirect linkages between cap-and-trade systems through a common credit system may not require such harmonization. As a result, in the near term, such indirect linkages may be easier to establish than some direct linkages.

- *In the near term, linkage may grow in importance as a core element of a bottom-up, de facto international policy architecture.* The European Union Emissions Trading Scheme has already established direct links with systems in neighboring countries, and the Clean Development Mechanism has emerged as a potential hub for indirect links among cap-and-trade systems worldwide. As new cap-and-trade systems appear in countries such as Australia, Canada, Japan, and the United States, the network of direct and indirect links will likely continue to spread.

- *In the longer term, linkage could play several roles.* A set of linkages, combined with unilateral emissions reduction commitments by many nations, could function as a stand-alone climate architecture. Such a system would be cost-effective, but might lack coordinating mechanisms necessary to achieve meaningful long-term environmental performance. Another possibility is that a collection of bottom-up links may evolve into a comprehensive, top-down agreement. In this scenario, linkages would provide short-term cost savings while serving as a natural starting point in negotiations leading to a top-down agreement. That agreement might continue to use linkage as a means of reducing abatement costs and improving market liquidity.

- *A post-2012 international climate agreement could include several elements that would facilitate future linkages.* Such an agreement could establish an agreed trajectory of emission caps or allowance prices, specify harmonized cost-containment measures, and establish a process for making future adjustments to key design elements. It could also create an international clearinghouse for transaction records and allowance auctions, provide for ongoing improvements to the CDM, and build capacity in developing countries. Such an agreement should avoid features that may adversely affect the performance of linkages, such as by encouraging strategic behavior or imposing "supplementarity" restrictions, which require countries to achieve some specified percentage of emissions reductions domestically.

Conclusion

In the near term, indirect linkages of cap-and-trade systems via a common emission-reduction-credit system could achieve meaningful cost savings and risk diversification without the need for much harmonization between systems. In the longer term, international negotiations could establish shared environmental and economic expectations that would serve as the basis for a broad set of direct links among cap-and-trade systems. This progression could promote near-term goals of participation and cost-effectiveness while helping to build the foundation for a more comprehensive future agreement.

Author affiliation

Judson Jaffe is a vice president at Analysis Group, Inc.
Robert N. Stavins is the Albert Pratt Professor of Business and Government at the Harvard Kennedy School.

Appendix 1.4 The case for charges on greenhouse gas emissions

Richard N. Cooper

Overview

This paper proposes a world-wide charge on emissions of greenhouse gases from all sources. The charge would be internationally adjusted from time to time, and each country would collect and keep the revenue it generates.

Discussion

Seriously addressing carbon dioxide emissions requires a worldwide approach, not one limited to today's rich countries. Levying a charge on CO_2 raises the price of CO_2-emitting activities, including fossil fuel use, and thus is the most direct method of influencing consumer and industrial behavior on a world-wide scale.

Compared to the alternative of a cap-and-trade (CAT) system, a carbon charge has two compelling advantages. First, under a CAT scheme, governments would need to allocate valuable emission permits to domestic firms or residents. This will foster rampant corruption in many countries. A universal CO_2 charge would avoid such problematic and politically indefensible transfers among countries. Second, it may be impossible to negotiate the meaningful global emission cap required for a CAT system. In contrast, a carbon charge would generate significant revenues that could be used to increase government spending or to reduce other charges. A portion of the revenues might also be used to finance climate-relevant research and development. Additionally, a carbon charge may be less objectionable to developing nations than an emissions cap. For example, a carbon charge is in complete harmony with China's official energy strategy.

The European Union seems committed to a cap-and-trade system: Can CAT systems and charge systems co-exist? The answer: Yes, provided several conditions are met. First, the trading prices under the CAT system should average no less than the internationally agreed carbon charge. Second, if the permit trading price fell below the agreed charge by some percent for more than a certain period of time, trading partners would be allowed to levy countervailing duties on their imports from the CAT countries. Third, countries could not provide charge rebates on their exports. Finally, CAT countries could not give away emission permits.

Key findings and recommendations

- *Since climate change is a global problem, geographic coverage of the carbon charge should be as broad as possible.* The initial scheme need not cover all countries. However, it should cover the three or four dozen countries that account for the vast majority of world emissions. The charge should cover all the significant greenhouse gases, insofar as is practical.
- *The level of the charge would be set by international agreement and would be subject to periodic review every five or ten years.* Initially, it should be high enough to affect behavior significantly, but not so high as to lead to unwarranted adjustments. A good starting price might be $15 per ton of CO_2-equivalent, but this would be determined by negotiation.
- *To minimize administrative costs, the charge should be assessed at upstream locations.* For example, the carbon content of oil should be charged at refineries, gas should be charged at major pipeline collection points, and coal should be charged at mineheads or rail and barge collection points. All but the poorest and least competent nations should have sufficient administrative capacity to administer such a plan, and those lacking institutional capacity are likely to be low emitters.
- *The treaty would include monitoring and enforcement measures.* The International Monetary Fund would monitor and support signatory nations' efforts at legislative and administrative compliance. Non-compliant or non-signatory countries could also be subject to countervailing duties. This possibility would provide a potent incentive for most countries to comply with an agreement.
- *Each country would retain the revenues it collected from the carbon charge and could use those revenues to reduce other charges or increase government expenditures.* The macroeconomic impact of the carbon charge could be kept low by introducing the charge gradually, at a pace consonant with the increased expenditures or reductions in other charges.

- *The revenues and economic impacts of a carbon charge would be substantial, but not overwhelming.* For example, in 2015, a charge of $15 per ton of CO_2 would generate approximately $515 billion in world-wide revenues, or about 0.7 percent of gross world product in that year. In the United States, this would add about 1.78 cents per kilowatt-hour to the cost of coal-generated electricity and 13 cents to a gallon of gasoline.

Conclusion

An international charge on greenhouse gas emissions would be an effective and feasible mechanism for dealing with global climate change.

Author affiliation

Richard N. Cooper is the Maurits C. Boas Professor of International Economics at Harvard University.

Appendix 1.5 Towards a global compact for managing climate change

Ramgopal Agarwala

Overview

Despite the dangers of climate change, there has been little progress toward a global climate agreement. This paper presents an approach that could reconcile the perspectives of developing and developed countries, differences which have deviled potential agreements for quite some time.

Discussion

The primary factor behind lack of progress in multilateral negotiations is the changing power equation in the global economy. Until recently, developed countries were the undisputed leaders in these negotiations. However, the global South has stood up and is determined to make its voice heard. In view of the serious risks that humanity faces from continued global warming, a failure to reach agreement on managing climate change would be most unfortunate.

Marking real progress towards an agreement will require being frank about the problems underlying climate negotiations. Both developed and developing nations must face up to some "inconvenient truths." First, developed countries must accept responsibility for their historic emissions of greenhouse gases. Second, if the western lifestyle is not replicable for the world as a whole, it must be modified in both developed and developing countries. Third, the global South needs in its own research to understand that climate change discussions are not a tool that the North is using to slow the economic and political rise of the South. Fourth, developing countries must stop hiding behind the poor. Fifth, the present discussions of climate change impacts concentrate too much on the very long term. Something more immediately relevant is needed.

Despite these challenges, a credible global compact is possible. It will need to satisfy five criteria. First, it has to be comprehensive by including both developed and developing countries. Second, it has to be equitable. Third, the targets on emissions have to be realistic. Fourth, the program has to be efficient. Fifth, the program has to develop an institutional mechanism for effective implementation.

Key findings and recommendations

- *The Kyoto Protocol satisfies none of the five criteria described above for a credible global compact.* First, the Protocol, which covered only 30 percent of global emissions in 2003, does not provide a comprehensive mechanism for emission control. Second, the Protocol's targets are based primarily on political bargaining, not on equity. Third, the Protocol does little to indicate, even in broad terms, the programs of technological dissemination, incentives, and resources needed for achieving the targets. Fourth, the Protocol's cap-and-trade system faces severe practical problems. Fifth, the Protocol relies on voluntary self-enforcement and allows countries to withdraw from the agreement without penalty. In view of these limitations, it is not surprising that the Protocol is not achieving its objective of reducing carbon emissions.

A post-Kyoto agreement should include the following features:
- *The agreement should set a realistic target of stabilizing global CO_2 emissions at 2003 levels until 2050 and reducing them by 50 percent by 2100.* If emissions are allocated on a per capita basis, this will require reduction in emissions in developed countries by about 70 percent by 2050, and allow an increase in developing countries' emissions by about 70 percent. This will allow 6 percent annual growth in GDP in developing countries provided they can reduce carbon intensity of their GDP by 5 percent per year.
- *The agreement should set appropriate carbon prices by eliminating subsidies to emitters (particularly energy subsidies) and establishing a carbon tax.* This approach would minimize bureaucratic intervention, provide appropriate incentives, and raise revenue for mitigation and adaptation.
- *The agreement should support development and dissemination of carbon-saving technologies.* The massive programs popularizing family planning in developing countries provide an example of what should be done for

climate change. More specific suggestions include patent buy-outs, reduction of tariffs on sale of technologies, a global clean-energy venture capital fund, transfer of technologies to public domain, licensing schemes with reduced duration of intellectual property rights, and flexible technology transfer mechanisms.

- *The agreement should be negotiated at the United Nations, but should be implemented using Bretton Woods Institutions, namely the International Monetary Fund (IMF) and the World Bank.* The UN system is the appropriate forum for negotiations and agreements on a global program for climate change. However, implementation of the agreements is likely to require substantial financial resources as well as formulation of concrete projects and programs. The IMF could be an ideal agency for reviewing the issues of carbon subsidies and carbon taxation at national and global levels. The World Bank could serve as an ideal agency for supporting projects and programs for carbon reduction.

- *Seigniorage (revenue from printing currency) from the creation of a new global currency to replace the dollar could be used to fund adaptation efforts.* It appears increasingly likely that global warming of more than two degrees centigrade may be unavoidable and that funding of adaptation efforts will be needed. However, developed nations may be reluctant to transfer large sums to developing nations. Revenues from printing and issuing a new global currency could be a good alternative source of funding available only to countries that comply with emissions targets.

Conclusion

The ideas in this paper are undoubtedly ambitious. However, with the fate of humanity at stake, the world needs nothing less.

Author affiliation

Ramgopal Agarwala is Distinguished Fellow, Research and Information System for Developing Countries, in New Delhi, India.

Appendix 1.6 Sectoral approaches to a post-Kyoto international climate policy framework

Akihiro Sawa

Overview

The Kyoto Protocol uses a top-down mechanism to negotiate economy-wide emissions caps. This paper proposes an alternative "sectoral" approach, which would determine industry-level emissions reduction targets based on technological analyses.

Discussion

A sectoral approach to a future climate agreement has a number of advantages over the existing Kyoto framework. First, a sectoral approach would encourage the involvement of a wider range of countries, since it would include incentives targeted at specific industries in those countries. Second, a sectoral approach would resolve a variety of issues concerning international competitiveness. Industries would make cross-border commitments to equitable targets, thereby mitigating concerns about unfair competition and emissions leakage (in which energy-intensive firms relocate to countries with weaker emissions regulations). Third, a sectoral approach may be more convincing to interest groups, since calculating emissions targets based on technological analysis may reduce uncertainty about future marginal costs of abatement. Fourth, a sectoral approach would achieve effective emissions reductions by promoting technology development and transfer.

A sectoral approach also has a number of weaknesses. First, it may be difficult to negotiate an international agreement based on a sectoral approach, since it would have large transaction costs, create uncertainty about the investments of countries that already participate in emissions trading schemes, and complicate negotiations by allowing countries to raise competitiveness issues not directly related to carbon restrictions. Second, a sectoral approach would reduce cost-effectiveness. Unlike an economy-wide cap-and-trade system, which can achieve cost-effectiveness by exploiting abatement opportunities with minimum costs, a sectoral approach would force reductions upon specific sectors. Third, a sectoral approach cannot achieve environmental effectiveness, since it does not induce mitigation actions from all sectors. Fourth, a sectoral approach entails a high level of government intervention, both by requiring additional government authority in collecting industry data and by requiring governments to set sector-specific regulations. Fifth, a sectoral approach faces challenges related to data collection and antitrust laws.

Key findings and recommendations

- *This paper proposes a "Policy-Based" sectoral approach for an international climate agreement that addresses some of the problems associated with previous sectoral designs.* Under this new approach, governments would negotiate national and sectoral emission targets and policies based on sectoral-level analyses of what is feasible given projected technological progress. For the purposes of the agreement, industries would be categorized into three major sectors. Group I sector includes energy-intensive industries that engage in significant international trade. Group II sectors include primarily-domestic industries, such as electricity and road transport, for which efficiency benchmarks and best practices can be identified easily. Group III sectors include households and commercial establishments for which standards would be difficult to set due to the wide range of activities and technologies involved.

- *The agreement would require industrialized countries to take on binding reduction targets at the national level and for Group I.* Developing countries' commitments would be non-binding and could include economic or sustainable development policies with only secondary effects on greenhouse gas emissions. The agreement would promote cost-effectiveness by establishing an intensity-based market for emissions trading.

- *Under this sectoral approach, governments would have discretion to choose the policies used for domestic implementation.* For example, the European Union might choose to meet its Group I commitments using domestic emissions trading. In Japan and China, Group I options might include establishing formal agreements with domestic industries or setting energy conservation standards. Potential Group II policies include financial support for technical support and technology transfer. In Group III, policies could include energy efficiency standards and tariff reductions. Over time, policy measures (particularly Group I policy measures) should converge towards a single framework, such as international linkage among national emissions trading systems.

- *The international agreement implementing the sectoral approach would include a variety of enforcement measures.* Nations that fail to comply with binding national or sectoral emissions targets would be required to purchase emissions credits from other countries or to pay additional penalties in the following commitment period. Trade sanctions could be used for enforcement, to mitigate international competitiveness concerns, and to encourage participation in the treaty.

- *To encourage widespread participation, the agreement would provide financial and technological incentives to developing countries.* For example, developing countries could be given "no-lose" targets that allow them to sell credits if they reduce their emissions intensity below a certain threshold. Alternatively, they could be given dual intensity targets that require them to meet a "compliance" target, but allow them to sell credits for emissions reductions above a higher "selling" target. Finally, industrialized countries should provide incentives for private firms to transfer technology to developing world projects.

Conclusion

A sectoral approach to a future climate agreement may help to solve some of the problems of the Kyoto Protocol. However, some issues related to a sectoral approach – including lower cost-effectiveness, the difficulty of data collection, and the complexity of sector-level negotiations – remain unresolved.

Author affiliation

Akihiro Sawa is Senior Executive Fellow at the 21st Century Public Policy Institute, in Tokyo, Japan.

Appendix 1.7 A portfolio system of climate treaties

Scott Barrett

Overview

Since the Kyoto Protocol has so far failed to achieve the core objective of reducing global greenhouse gas emissions, this paper proposes a radically different approach. Rather than attempting to address all sectors and all types of greenhouses gases under one unified treaty, the author argues for a system of linked international agreements that separately address different sectors and gases, as well as key issues like adaptation and technology R&D, and last-resort remedies like geoengineering.

Discussion

The failure of the Kyoto Protocol (emissions are still rising) can be traced to a lack of enforcement. Since sovereign nations cannot be compelled to act against their wishes, successful treaties must create adequate incentives for participation and compliance. The Kyoto Protocol lacks both.

One much-discussed remedy is trade restrictions (especially, a border tax), which would financially penalize countries that refuse to join a climate agreement. But this approach is problematic. First, trade restrictions would need to be both credible and severe – two characteristics that are often in tension. Second, their legitimacy may be open to challenge, and their use could spur retaliatory measures. Finally, trade restrictions would also have to be used to enforce compliance, and it is not clear that parties to a future treaty would agree to this – particularly when many countries are falling short of meeting their existing commitments.

If economy-wide obligations cannot be enforced by trade restrictions or any other means, perhaps a different approach should be tried – one that focuses on individual sectors. The Kyoto Protocol itself treats some sectors separately, notably marine transport and aviation. A sectoral approach also has the advantage that, if policies designed for a given sector prove ineffective, their failure need not drag down the entire enterprise. Similar arguments can be made for separate approaches to different types of greenhouse gases.

Key findings and recommendations

The paper goes on to discuss specific aspects of a portfolio approach to reducing global climate-change risks:

- *Sector-level agreements should provide global standards for specific sectors or categories of greenhouse gas sources (for example, the aluminum industry).* Developing countries should not be exempted from these standards but should be offered financial aid to help them comply. Finally, trade restrictions should be used to enforce agreements governing trade-sensitive sectors (i.e., aluminum), where such sanctions can be both effective and credible.
- *R&D obligations should be linked to emission reduction policies.* For example, an agreement could require that all new coal-fired power stations be fitted with carbon capture and storage, with this obligation being binding only so long as the treaty met minimum participation requirements. Such an agreement would reduce incentives for free-riding and spur R&D in an area where countries might otherwise be likely to under-invest. This approach would also address a key shortcoming of the Kyoto Protocol – its failure to directly stimulate R&D investments.
- *Adaptation assistance to developing countries should be provided, consistent with the obligation already articulated in Article 3 of the United Nations Framework Convention on Climate Change.* All nations have strong incentives to adapt, but only rich countries have the resources and capabilities to insure against the consequences of climate change. In fact, rich countries may be tempted to substitute investments in adaptation (the benefits of which can be appropriated locally) for investments in mitigation (the benefits of which are distributed globally). If so, this would leave developing countries even more exposed to climate risks and tend to widen existing disparities. It is not yet clear what form a new approach to adaptation assistance would take, but it is possible to identify several critical areas for investment, including agriculture and tropical medicine.
- *Geoengineering and air capture have a role to play in the portfolio of options.* Geoengineering strategies attempt to limit warming by reducing the amount of solar radiation that reaches the Earth's surface (the most

commonly discussed approach involves throwing particles into the atmosphere to scatter sunlight). Because this form of geoengineering could be implemented relatively cheaply, the greater challenge may be to prevent nations from resorting to it too quickly or over other countries' objections. Air capture refers to strategies for removing carbon from the atmosphere – possible options include fertilizing iron-limited regions of the oceans to stimulate phytoplankton blooms or using a chemical sorbent to remove carbon directly from the air. The latter approach would be very costly and is unlikely to be implemented unilaterally. The proposal encourages more R&D for these efforts and global negotiation over whether and when they should be used.

Conclusion

The proposed multi-track climate treaty system is not perfect but could offer important advantages over the current approach. By avoiding the enforcement problems of an aggregate approach and by taking a broader view of risk reduction, the portfolio approach provides a more effective and flexible response to the long-term global challenge posed by climate change.

Author affiliation

Scott Barrett is Lenfest Earth Institute Professor of Natural Resource Economics, Columbia University.

Section II

Negotiation, Assessment, and Compliance

Appendix 1.8 How to negotiate and update climate agreements

Bård Harstad

Overview

The outcome of negotiations depends on the bargaining rules. Those hoping for a successful international climate agreement should thus pay attention to the rules governing the negotiation process. This paper describes several bargaining rules that would facilitate agreement on a post-2012 climate treaty.

Discussion

Prepare for negotiations. Not only is it challenging to negotiate a climate treaty from scratch, but climate change is a dynamic problem; we are going to learn more about the benefits and costs; and therefore any treaty will have to be updated and renegotiated later on.

Anticipating future negotiations generates what are called "hold-up problems" in economics: countries that have poor abatement technology are able to hold up the countries with better technologies and require that the high-tech countries share their technologies or contribute the most to the next commitment period. Anticipating this, each country is discouraged from investing in abatement technology. Moreover, to enhance its future bargaining position, a country may want to (i) adapt more to climate change than socially optimal, (ii) signal reluctance to negotiate by delay, or (iii) delegate bargaining authority to representatives that are less favorable toward an agreement. Such strategies improve the bargaining position of an individual country, but collectively it becomes harder to arrive at an efficient outcome. If no rules are governing the negotiation process, these strategies can be extremely detrimental, and the gains from international negotiations shrink. It is thus immensely important to think carefully about the rules governing the negotiation process.

Many international agreements are governed by bargaining rules. Examples include the United Nations' voting rules and the World Trade Organization's "reciprocity principle." For climate negotiations, this paper proposes five bargaining rules that can mitigate the problems described above. The rules relate to the voting process, the use of harmonization or formulas, the time horizon of an agreement, the minimum number of participants and the default outcome (i.e., the outcome if the negotiations should fail). The rules may fruitfully be complemented by a connection to trade agreements.

Key findings and recommendations

- *Harmonization or formulas should be used to calculate national obligations and contributions.* If the distribution of contributions is determined by a formula, it is harder for a country to influence its own share of the burden. Enhancing its bargaining position is then less useful, and investments in R&D increase. Harmonization can be harmful if countries are heterogeneous, but formulas can be cleverly designed (to depend on GDP and growth, for example) to mitigate these concerns.
- *Climate treaties should have a long time horizon.* A longer time horizon reduces the frequency at which the agreement is renegotiated and thus countries' motives to enhance their bargaining power.
- *The unanimity requirement should be replaced by a majority or a super-majority rule when it comes to treaty amendments.* Unanimity means that even the most reluctant country must agree, and it is exactly this requirement that induces countries to strengthen their bargaining power. Reducing the majority requirement mitigates strategic considerations, and investments in R&D will increase.
- *A "minimum participation rule" can discourage free-riding.* If the treaty enters into force only after a certain number of countries have ratified it, opting out becomes less tempting if that could make the agreement unravel.
- *The treaty must specify the default outcome if the (re)negotiation process should break down, and this default outcome should be an ambitious agreement.* If the current climate negotiations fail, the outcome is no agreement at all. If the Doha trade round negotiations should fail, however, the outcome would be the existing set of trade agreements. The argument in this paper shows that the latter approach is better. Also, for environmental

agreements, the default should be an ambitious agreement rather than no agreement.

- *Investments in R&D, or trade in abatement technology, should be subsidized internationally.* This follows since countries may otherwise under-invest in R&D, as explained above.

- *Each rule is more important if the other rules are not followed.* This means that subsidizing R&D is more important if the time horizon is short and if formulas are not used in the negotiations. Similarly, the time horizon should be longer if unanimity is required for each amendment, or if the default outcome (if the renegotiations should fail) is no agreement rather than an ambitious agreement.

- *A linkage to international trade agreements makes each of the rules more credible and efficient.* For each rule above, a problem arises if a country can credibly threaten to opt out of the agreement unless it gets a more favorable deal. Opting out would be less tempting if a trade agreement provided additional benefits to the participating members. Thus, a linkage to trade benefits/sanctions is a "strategic complement" to each of the rules above.

Conclusion

Climate change agreements should – and certainly will – be updated over time. Anticipating future negotiations, countries may try to enhance their future bargaining power (e.g., by under-investing in R&D). This reduces the gain from international cooperation, unless we pay careful attention to how the bargaining process should be structured. This paper discusses several useful rules and how they relate to each other.

Author affiliation

Bård Harstad is Associate Professor of Economics at Northwestern University's Kellogg School of Management.

Appendix 1.9 Metrics for evaluating policy commitments in a fragmented world: the challenges of equity and integrity

Carolyn Fischer and Richard D. Morgenstern

Overview

Development of effective strategies to address climate change will require collective effort on the part of many countries over an extended period and across a range of activities. The challenge for the international community will be to judge the equity and integrity of the various national commitments.

Discussion

Since diverse actions by different nations seem an unavoidable part of future climate policy, a useful input to international negotiations would be some means of talking in a coherent and widely accepted fashion about what individual nations are doing to help reduce climate risk. This paper explores various metrics for evaluating the equity and integrity of individual nations' policy commitments.

Equity is a major concern because international climate negotiations are fundamentally about sharing a burden. There is a strong desire to compare efforts and assess whether countries are contributing their fair shares. Yet, comparing efforts involves two kinds of exercises, neither of which lends itself to clear and fair metrics. The first exercise is to take a portfolio of disparate national policies and compare them according to some consistent measure that reflects effort, cost burden, or emissions reductions. The second exercise is to place that measure of effort for each country into an appropriate context reflecting its socio-economic and other circumstances, in order to assess fairness. The basic problem is that clear metrics are not always fair, and fair metrics are not always clear. This paper considers a number of alternative approaches to measuring climate policy contributions, including measures of emissions performance, reductions, and costs. Each can provide some valuable information, but none is terribly satisfying as a reliable measure of effort or equity.

Evaluating the integrity of a collection of country commitments also requires two levels of analysis. First, the credibility of the commitments must be assessed; that is, will the countries actually undertake the measures, and can they be monitored and verified? This question requires the ability to conduct *ex post* analysis to support enforcing the agreement and engaging in commitments. The second level of analysis related to integrity involves assessing whether the proposed effects of the commitments are themselves credible; that is, do we reasonably expect the set of policies being undertaken to lead to the stated emissions goals?

Key findings and recommendations

- *There is a clear need to improve the current reporting system in order to provide greater confidence to negotiators about the credibility of countries' activities.* Most importantly, reported activities need to be presented in a more uniform, consistent fashion. The breadth of the different reporting practices currently used can mask genuine differences among countries. A first order of business should be the development of a much tighter, narrowly-defined set of reporting guidelines designed to reflect genuine differences in activities among nations.
- *Consensus on a particular metric for indicating equitable burden sharing is likely to be elusive.* Each country has its own incentive to choose measures of effort by which it is likely to perform relatively well. Some metrics are straightforward to calculate, and they are somewhat informative, although imperfect indicators of burden. Other metrics are unlikely to be reported reliably. One metric has the advantage of indicating the cost-effectiveness of the international distribution of effort, and that is marginal abatement costs. It is also an important indicator of the controversial competitiveness impacts of climate policies vis-à-vis trading partners.
- *For* ex post *verification, the simplicity of an aggregate, economy-wide emissions target, or even one expressed as emissions intensity, is appealing.* Existing data and reporting systems are certainly compatible with an aggregate approach. When subnational or specific regulatory or voluntary programs are used, descriptive, institutionally-oriented information must be supplemented with detailed data on the actual implementation and

performance of these measures. Expressing the commitment goal in a way that focuses on subsector aggregates (such as total auto sector emissions) rather than reductions attributed to policies will allow for more direct comparison and verification of effectiveness.

- *Assessment of the integrity of* ex ante *commitments is the most important but also the most challenging area, because it requires modeling of counter-factuals.* The main focus should be on greater transparency in models and data, and greater standardization in methodologies to improve the consistency of analysis across sectors, policies, and countries. Another priority is the strengthening of UNFCCC peer reviews, which currently are not sufficiently rigorous to provide credibility to the negotiations.

- *While the multilateral trading system offers some lessons in negotiating and supporting international agreements, the circumstances are quite different for a climate framework.* In trade, countries negotiate the removal of barriers to foreign goods in exchange for the benefits of greater access to foreign markets. In climate change there is no such exchange; the negotiations are to share a global burden, from which the benefits are far removed in time and not excluded from non-members. Perceptions of fairness and effort thus play a greater role. National governments may not provide the objective evaluation that is essential to the serious comparison of national mitigation proposals. A greater role may need to be played by independent institutions, international organizations, academic researchers, and other third-party groups in strengthening the evaluation efforts that support the negotiations, and in integrating the evaluations into the full negotiation process.

Conclusion

No single metric can adequately address the complex issues of equity and integrity central to international agreement on climate change mitigation. Development of a common, consistent, and credible set of indicators should be prioritized to build the foundation of trust and transparency needed to underpin multifaceted commitments.

Author affiliations

Carolyn Fischer is a Senior Fellow at Resources for the Future.
Richard D. Morgenstern is a Senior Fellow at Resources for the Future.

Justice and climate change

Eric A. Posner and Cass R. Sunstein

Overview

Climate change raises difficult issues of justice, particularly with respect to the distribution of burdens and benefits among poor and wealthy nations. To illuminate these issues, this paper focuses on the narrower question of how to allocate greenhouse gas emission rights within a future international cap-and-trade system. In particular, it highlights shortcomings in an approach that is often advanced on fairness grounds: a per capita allocation, in which emissions permits are distributed to nations on the basis of population.

Discussion

In an international cap-and-trade system, participating nations would agree to an overall "cap" on emissions, and each nation would receive a share of total emission rights or permits under the cap. The allocation of emission rights would not bear on the effectiveness or efficiency of the program, but it would have important distributive impacts.

The two approaches most commonly considered for allocation are based on existing emissions and population. A pragmatic argument can be made for the former on grounds that it would more likely win the support of wealthy and powerful nations. However, this approach also violates basic notions of fairness and could be seen as limiting the economic development rights of poor nations. By contrast, an allocation that awards equal emission rights on a per capita basis strikes many observers as inherently more equitable and responsive to welfare concerns. Under this approach, countries with high per capita emissions (such as the United States) would lose relative to countries with large populations but low per capita emissions (such as China).

The intuitive appeal of a population-based allocation may tend to obscure its practical and theoretical weaknesses. Close examination of this approach in light of several complicating factors – including the lack of correlation between population and wealth, the differential benefits of climate-change abatement in different countries, and the realities of governance in many poor nations – reveals potent shortcomings. Understanding these shortcomings may shed light on better approaches to reconciling climate change and justice concerns.

Key findings and recommendations

- *A variety of approaches could be used to allocate emission rights among nations. Allocations based on population or on redistributing wealth are generally more equitable than allocations that award permits on the basis of current emissions.* In choosing among approaches, however, feasibility must also be considered. Poor nations that are especially vulnerable to climate change may have the most to lose if insistence on principle precludes agreement in practice.
- *A per capita allocation, while preferable to an emissions-based allocation, would not in practice satisfy objectives of fairness and welfare redistribution.* The argument that every individual person should be entitled to the same emissions rights seems intuitively compelling, but it fails along several dimensions when one considers the following:
  - Not all large countries are poor and not all small countries are rich. A per capita allocation would give some rich and populous countries a large share of emission rights. Conversely, while populous countries with low per capita emissions (like China and India) would certainly benefit, some even poorer countries with small populations would lose.
  - A complete assessment of the impacts of an international policy must take into account the benefits of climate change abatement, as well as the revenue effects of permit allocation. If the aim is to equalize impacts, not just permit endowments, the different environmental benefits that accrue to different countries under the policy must also be considered. A citizen of Russia, for example, can expect to benefit less from greenhouse-gas reductions than a citizen of India (because warming is likely to be far more negative for India than for Russia). If both get an equal allotment of permits, the net result is not 'fair' because the Indian citizen benefits more from the policy than the Russian citizen.

- Transferring wealth in the form of emission rights to the governments of poor countries does not mean that this wealth will be fairly distributed to citizens of those countries. Just as foreign aid is often misused or diverted, a per capita allocation could fail to achieve intended redistributive effects, especially in countries with corrupt or ineffective governments.
- *If the goal is a more equal distribution of wealth, an approach that is openly redistributive is better than a per capita allocation.* From a welfarist standpoint, in which the aim is to maximize global welfare and minimize inequality, an approach that has wealthy nations pay poor ones for emissions reductions or for adaptation – or both – is superior to a per capita allocation.

Conclusion

While a per capita approach to the international allocation of emissions rights is often advanced on welfare and fairness grounds, neither principle is well served by this approach. Alternatives that target assistance to poor nations – and more particularly, to poor people in poor nations – must still meet the test of feasibility, but would succeed far better than many current proposals in reconciling the aims of climate policy with the claims of distributive justice.

Author affiliation

Eric A. Posner is the Kirkland & Ellis Professor of Law at the University of Chicago Law School.
Cass R. Sunstein is the Felix Frankfurter Professor of Law at Harvard Law School.

Appendix 1.11 Toward a post-Kyoto climate change architecture: a political analysis

Robert O. Keohane and Kal Raustiala

Overview

A successful climate change regime must secure sufficient participation, achieve agreement on meaningful rules, and establish effective mechanisms for compliance. This chapter focuses on the problem of compliance, arguing that contrary to current provisions in the Kyoto Protocol, only a system of buyer liability (rather than seller or hybrid liability) is consistent with existing political realities. Drawing analogies to international bond markets, we propose a system of buyer liability that would endogenously generate market arrangements, such as rating agencies and fluctuations in the price of emissions permits according to perceived risk. These features would in turn create incentives for compliance with national emissions caps without resorting to unrealistic and ineffective inter-state punishments.

Discussion

In world politics, states must prefer participation to non-participation for an international regime to be viable. In this context, a cap-and-trade approach offers many advantages beyond efficiency, since it creates opportunities to offer disguised subsidies to less motivated states in the form of excess emissions permits that can be profitably sold internationally. But one problem with such an approach is that, since external enforcement is generally ineffective in international relations, over-selling of permits by such states is difficult to deter. Anticipating this result, states likely to be net buyers will tend to refuse to join the regime. The problem of compliance must therefore be solved, or at least alleviated, in order to both ensure participation and make emissions reductions meaningful.

Entities in advanced industrial democracies are likely to be the primary buyers of permits in any international cap-and-trade system. As in international bond markets, in which buyers bear the cost of fluctuations in the value of bonds, we propose a system in which annual permit prices would depend on market expectations of their validity, which would in turn depend both on the reputation of the sellers and on information about their validity gained from *ex post* assessments in previous years. Buyers, who generally will face well-functioning national regulatory systems, will have incentives to seek information about the quality of permits, and ratings agencies will likely emerge to perform this function. Making buyers liable for the validity of the permits they purchase thus puts the onus for compliance on those residing in the states generally most committed to the regime, rather than those generally least committed.

Under a buyer liability regime, penalties for exceeding national emissions caps would be imposed on sellers in the form of lower permit prices, rather than through an (ineffective) external enforcement mechanism. Permit validity would be assessed on a "jurisdiction-equal" and annual basis: that is, all permits from a given jurisdiction in a given year would be discounted by the same factor in the event that emissions in the selling jurisdiction rose beyond the mandated cap. This determination would be made at the end of each year by a centralized body tasked with assessment. This system provides incentives for seller governments to ensure overall permit quality in order to achieve the highest price, and it incentivizes buyers to evaluate and assess permit validity and to price them accordingly.

Key findings and recommendations

- *An international cap-and-trade regime must encompass developing as well as developed states.* A regime that awards developing states with excess emissions permits is most likely to reconcile the need for meaningful reductions and adequate participation.
- *Making buyers liable for the validity of emissions permits has compelling advantages.* Permit buyers are likely, at least initially, to be concentrated in developed countries that have the institutional capacity and political accountability to enforce program requirements and make up for permit shortfalls. This approach also gives sellers strong economic incentives to maintain permit quality so as to maximize the monetary value of these tradable assets.

- *Any discounting of permits to reflect quality problems should be assessed on a "jurisdiction-equal" basis;* that is, all permits from a given country would be equally affected, creating strong internal pressures for governments to provide necessary monitoring and enforcement.
- *A prompt and credible system for assessing permit quality is critically important.* Such a system must be technically feasible and minimize opportunities for strategic manipulation. A key virtue of a buyer liability system is that it separates assessment (done by a centralized body) from enforcement (decentralized through markets).

Conclusion

A cap-and-trade system with buyer liability has important political and pragmatic advantages. By building on the preferences of domestic publics in democratic states to induce participation by less motivated countries and by giving buyers and sellers compatible incentives for maintaining permit quality, such a system provides the most promising basis for a workable system of emissions reduction.

Author affiliation

Robert O. Keohane is Professor of International Affairs at Princeton University.
Kal Raustiala is a professor in the School of Law at the University of California at Los Angeles, and at the UCLA International Institute.

Section III

The Role and Means of Technology Transfer

Appendix 1.12 International climate technology strategies

Richard G. Newell

Overview

Policies facilitating innovation and large-scale adoption of low-carbon technologies could play an important role in global efforts to address climate change, alongside policies targeted directly at reducing emissions. This paper considers opportunities for improved and expanded international technology development and transfer strategies within the broader context of international agreements and institutions for climate, energy, trade, development, and intellectual property.

Discussion

Achieving the deep reductions in greenhouse gas (GHG) emissions necessary to stabilize atmospheric GHG concentrations will require substituting energy technologies with low to zero net GHG emissions throughout the global energy system. The scale of the task is immense. According to the secretariat of the United Nations Framework Convention on Climate Change (UNFCCC), an additional $200 billion in global investment and financial flows will be required annually by 2030 just to return GHG emissions to current levels. This is in addition to the roughly $900 billion per year in global energy infrastructure investments the International Energy Agency (IEA) estimates will be needed over the 2006–2030 timeframe simply to keep up with expected demand growth.

In this context, innovations that significantly reduce the cost disadvantage of climate-friendly technologies relative to the competition could provide enormous economic and environmental benefits, substantially reducing the costs of mitigation and potentially making it feasible to undertake more significant reductions. The associated policy debate is therefore not so much over the importance of technology per se, but rather over what policies and institutions would most effectively and efficiently spur the technology advances needed to address the problem.

Key findings and recommendations

Recognizing that any successful effort to accelerate and then sustain the rate of development and transfer of GHG mitigation technologies must harness a diverse set of markets and institutions, this paper recommends several specific issues and actions for consideration in international climate policy discussions:

- *Long-term national commitments and policies for emission mitigation are crucial to providing the necessary private sector incentives for technology development and transfer.* Absent the market pull of a financial incentive for reducing GHG emissions, R&D efforts by themselves will have little impact. Private sector trade, investment, and innovation – motivated by widespread global demand for low-GHG technologies – will be essential to moving the energy system in the desired direction.
- *Financial assistance to developing countries for technology transfer and capacity building is also necessary.* At the same time, GHG-related guidelines for financing by Export Credit Agencies (ECAs) and multilateral development banks can help ensure that trade and development assistance investments are consistent with climate mitigation goals.
- *In addition to increased incentives, barriers to climate-friendly technology transfer could be reduced through a World Trade Organization (WTO) agreement to reduce tariff and non-tariff barriers to trade in environmental goods and services.* Development and harmonization of technical standards – by international standards organizations in consultation with the IEA and WTO – could further reduce impediments to technology transfer and accelerate the development and adoption of climate-friendly innovations.
- *To support the upstream supply and transfer of new climate technology innovations internationally, strategies are needed to increase and more effectively coordinate public funding of R&D.* Specifically, a framework for coordinating and augmenting climate technology R&D could be organized through a UNFCCC Expert Group on Technology Development, supported by the IEA. Broadening IEA participation to include large non-OECD energy consumers and producers could also facilitate such coordination. An agreement could include a process for reviewing country submissions

on technology development and for identifying redundancies, gaps, and opportunities for closer collaboration. A fund for cost-shared R&D tasks and international technology prizes could be established to provide financing for science and innovation objectives that are best pursued in a joint fashion. The agreement could also include explicit targets for increased domestic R&D spending on GHG mitigation.

- *Strategies are also needed to resolve impediments to knowledge transfer in the context of policies for the protection of intellectual property.* To that end, an Expert Group on Technology Development, the World Intellectual Property Organization (WIPO), and the WTO could work jointly to develop recommendations for addressing technology development and transfer opportunities and addressing intellectual property issues. A fund could be established in WIPO or another appropriate body for related technical assistance, capacity building, and possibly to purchase intellectual property or cover related costs.

Conclusion

The range of opportunities for improved and expanded international climate technology development and transfer extends well beyond the usual boundaries of environmental decision makers to the broader context of international agreements and institutions for energy, trade, development, and intellectual property. The technology challenge implicit in successfully addressing climate change along with our other energy problems is enormous. It requires a portfolio of strategies for reducing barriers and increasing incentives for innovation across international agreements and institutions in a way that maximizes the impact of scarce public resources and effectively engages the capacities of the private sector.

Author affiliation

Richard G. Newell is the Gendell Associate Professor of Energy and Environmental Economics at Duke University's Nicholas School of the Environment.

Mitigation through resource transfers to developing countries: expanding greenhouse gas offsets

Andrew Keeler and Alexander Thompson

Overview

Developing country participation is an increasingly contentious issue – and a potential deal-breaker – in current efforts to forge new international agreements for addressing climate change. Emissions offsets, which allow rich countries to finance mitigation actions in developing countries, may offer a partial solution to the current impasse. This paper proposes a more expansive approach to offsets that would meet the different objectives of industrialized and developing countries while providing substantial support for long-term investments and policy changes to reduce greenhouse gas emissions in the developing world.

Discussion

Developing and industrialized countries share a common view about climate change: each thinks the other should be doing more to solve the problem. Rich countries point to the rapidly growing contribution of the developing world, which now accounts for more than half of global greenhouse gas emissions, while poorer countries point to the North's responsibility for historic emissions and emphasize their own obligation to attend first to urgent development needs. There is no doubt that effective mitigation strategies will require developing country actions sooner rather than later. For both political and economic reasons, however, developing countries are highly unlikely to accept binding emissions targets in any upcoming evolution of an international climate regime.

This paper argues that emission offsets offer a financially and politically viable path toward reconciling the divergent interests of industrialized and developing countries. Offsets can help industrialized countries meet short- and long-term mitigation objectives while simultaneously providing a mechanism for transferring resources to the South. Realizing this potential requires changes to the existing Clean Development Mechanism (CDM). In particular, less emphasis on strict ton-for-ton accounting and more emphasis on a range of activities and policy inputs that could produce significant long-term benefits might allow the international debate to move forward in more productive directions.

Key findings and recommendations

- *Change the criteria for offsets from "real, verifiable, and permanent reductions" to "actions that create real progress in developing countries toward mitigation and adaptation."* Strict, project-based accounting rules, while designed to protect the environmental integrity of trading programs, have increased transaction costs and limited the utility of the CDM mechanism. Developing country actions are more important than the sanctity of short-term targets in making progress on mitigating climate change risk. An approach toward developing countries that focuses less on specific tons and artificial targets, and more on evidence that positive and productive steps are being taken in exchange for resources expended, is justified.
- *Make a significant share of industrialized country commitments (whether international or domestic) achievable through offset payments to developing countries.* If industrialized countries aimed to purchase offset credits equivalent to at least 10 percent of their overall emissions targets, it would greatly expand the flow of resources available to support developing country actions.
- *Sell a specified portion of offset credits (perhaps 50 percent) up front and put the proceeds in a fund that would make investments in an array of projects throughout the developing world.* By allowing greater flexibility to support large-scale or non-standard projects, this approach could increase the geographic diversity of mitigation activities and reduce transaction costs. Some *ex ante* possibility exists that not all fund investments would deliver expected results, but this risk could be managed by viewing funded actions collectively as a portfolio and by implementing *ex post* adjustments in cases of severe non-performance or poor management.

- *Focus international negotiations on specific guidelines for an international offsets program.* Key issues include criteria for eligible activities, policies, and investments; requirements for documentation or accountability; mechanisms for *ex post* adjustment; criteria for the distribution of funds; and set-asides, if any, for particular types of projects or technologies. All options should be on the table, and negotiators should be willing to take risks and to accept the possibility that some actions might fail or only partially succeed.
- *Negotiations should delegate clearly delineated tasks to new and existing institutions for the purpose of managing and safeguarding the offsets program consistent with negotiated guidelines.* Specifically, the authors recommend giving the World Bank and the Climate Secretariat in Bonn lead roles in program implementation and information-sharing, respectively. In addition, they propose a ten-member committee composed of parties to the United Nations Framework Convention on Climate Change to oversee the new offsets program and an informal contact group of international organization officials to address potential conflicts with other international programs or policies.

Conclusion

International agreements and institutions are more likely to succeed if they are responsive to the needs and self-interest of individual countries. Given the importance of global cooperation to address climate change, a rigid focus on ensuring that offset policies do not undermine the sanctity of short-term targets – especially if it reduces the prospects for developing country participation – is misplaced. This proposal recognizes that poor countries will do significantly more if resources are available and that industrialized countries require real mitigation and some accountability for the resources they provide. It is based on the contention that offsets, as a mechanism for making sustained resource transfers to developing countries, are likely to be more politically feasible than other funding options in the near term. Overall, the proposal takes into account the distinct interests of both industrialized and developing countries by emphasizing environmental and development needs simultaneously.

Author affiliation

Andrew Keeler teaches at the John Glenn School of Public Affairs at the Ohio State University.

Alexander Thompson is Associate Professor of Political Science at the Ohio State University.

Possible development of a technology clean development mechanism in the post-2012 regime

Fei Teng, Wenying Chen, and Jiankun He

Overview

Many technologies that could mitigate greenhouse gas (GHG) emissions do exist, but not in developing countries. Thus, transfer of climate-friendly technologies from developed to developing countries is vital to solve the global climate challenge. This paper proposes an enhanced Clean Development Mechanism (CDM) regime with greater emphasis on technology transfer.

Discussion

The Kyoto Protocol's CDM was intended to achieve two objectives: to help industrialized countries to meet their emission targets in a cost-effective way and to support developing countries in achieving the goal of sustainable development. However, in its current form, the CDM does not promote large-scale technology transfer. This is problematic because reduction of greenhouse gases is highly dependent on the timing and scale of the introduction of new technologies. Without an innovative technology transfer mechanism, a huge amount of energy infrastructure in developing countries may be "locked-in" to a carbon-intensive mode.

There are a number of existing proposals for ways to improve the CDM in a post-2012 climate agreement. Under "Programmatic CDM", a public or private organization would coordinate and receive credits for a number of small-scale projects distributed over space, time, and owners. Under "Policy CDM", nations would receive credits for implementing policies and measures that are additional. Under "Sectoral CDM", businesses or governments would receive credits for reducing sectoral emissions below some pre-established industry-level emissions standard. These three alternatives focus on scaling up the CDM market, thereby increasing financial flows in carbon markets. However, the objective of a future CDM should not be picking low hanging fruit, but spurring new and replicable technology transfer from developed to developing countries. Neither the current CDM regime nor these alternative CDM regimes are sufficient to induce new and replicable technology transfer.

This paper proposes a new idea for a technology-oriented CDM. This "Technology CDM" would take technology transfer as the emissions-reducing activity for which credits are awarded. Technology CDM would have three unique features. First, the technology transfer goal must be clearly identified before any activities are implemented. The counterfactual base-line against which transfer goals would be evaluated would be defined by the assumption that new technologies would only be adopted in developing countries after a substantial delay. Second, only projects using the technology transferred under the program can receive credit for emissions reductions. Third, credits may be shared by the technology provider or the government of the host country if they provide enabling support for technology transfer and discounted or free licensing.

Key findings and recommendations

- *Technology CDM offers the opportunity to strengthen the technology trans-fer effects of the CDM in the near term without redesigning the whole system.* First, Technology CDM provides incentives that specifically address the problem of GHGs' intensive technology "lock-in" effects. The program would provide emissions credits only for transfers that speed the introduc-tion of new technologies. Second, the "additionality" of projects is rela-tively easy to prove, since the transfer of best available technology is always impossible in the absence of additional financial support. Third, by bun-dling together projects that use similar technologies, Technology CDM would create economies of scale and increase the likelihood of successful technology transfer.
- *Technology CDM also shares some of the key attractions of the existing CDM program.* First, it could lead to a reduction in GHG emissions com-pared to the baseline emissions that would occur in the absence of the program. Second, all projects that adopt one type of technology could use

the same baseline and the same monitoring methodology. Such inclusion would greatly simplify the whole process and reduce the transaction costs and registration risks.

- *Technology CDM fulfills the "measurable, reportable, and verifiable" requirement of the Bali Action Plan.* The emissions credits generated by Technology CDM can be regarded as a metric for technological and financial support provided to developing countries by developed countries. Additionally, the process would include documentation, validation, and verification.

- *Technology CDM reduces the risk of low-carbon investment. Under regular CDM, investors require a high-risk premium for financing unregistered emission units, as the probability of successful registration is uncertain.* However, under Technology CDM, once a technology is proved to be eligible, projects using the technology will be automatically accepted. Given the low risk in future flows of certified emissions reduction credits (CERs), project owners could sell their credits to raise capital before the operation of the project. Additionally, the way credits are awarded in Technology CDM gives some guarantee to the intellectual property rights of technology providers.

- *Technology CDM should focus not only on the final stage of CERs acquisition, but the whole technology transfer process.* This includes a number of steps: (1) defining technology transfer priorities; (2) establishing partnerships between public and private stakeholders; (3) addressing concerns of both technology providers and recipients; (4) bundling similar projects to achieve economies of scale; and (5) bundling similar projects to reduce transaction costs and further offset project costs.

Conclusion

The focus of this paper is not to design a new and comprehensive solution for the post-2012 climate regime, but to try to improve the existing CDM regime. The experience of past international negotiations indicates that developing a climate agreement will be an evolutionary and path-dependent process. A breakthrough idea is needed, but it should be arrived at through a series of gradual changes.

Author affiliation

Fei Teng is Associate Professor in the Institute of Energy, Environment and Economy at Tsinghua University, Beijing, China.

Wenying Chen is Professor in the Institute of Energy, Environment and Economy at Tsinghua University, Beijing, China.

Jiankun He is Professor and Director of the Low Carbon Energy Laboratory at Tsinghua University, Beijing, China.

Section IV

Global Climate Policy and International Trade

Appendix 1.15 Global environment and trade policy

Jeffrey Frankel

Overview

Global efforts to address climate change may be on a collision course with global efforts to reduce barriers to trade. This paper discusses the broad question of whether environmental goals in general are threatened by free trade and the WTO, before turning to the narrower question of whether trade policies likely to be included in various national efforts to address climate change are likely to come into conflict with WTO rules.

Discussion

With different countries likely to undertake different levels of climate-change mitigation, the concern arises that carbon-intensive goods or production processes could shift to countries that do not regulate greenhouse gas (GHG) emissions. This so-called "leakage" phenomenon is viewed as problematic – by environmentalists because it would undermine emission reduction objectives and by industry leaders and labor unions because it could make domestic products less competitive with imports from nations with weaker GHG regulations. Thus, various trade measures – including provisions for possible penalties against imports from countries viewed as non-participants – are increasingly being included in major climate policy proposals in the United States and Europe.

These concerns represent the latest and most prominent manifestation of a broader set of fears about the impacts of free trade and globalization. Do favorable "gains from trade" – including the environmental improvements that sometimes come with economic growth and the benefits of greater openness and technology transfer – outweigh the potential for adverse impacts, if free trade spurs countries and firms to seek competitive advantage through lower environmental standards? A number of studies have found that the impacts of trade on pollution are more beneficial than detrimental, though this is not true for CO_2 emissions.

In any case, the widespread impression that the WTO is hostile to environmental concerns seems to have little basis in fact. The WTO's founding articles cite environmental protection as an objective; environmental concerns are also explicitly recognized in several WTO agreements. A review of recent WTO rulings finds support for the principle that countries not only have the right to ban or tax harmful products, but that trade measures can also be used to target processes and production methods, provided they do not discriminate against foreign producers. The question, then, is how to address concerns about leakage and competitiveness in a way that does not run afoul of WTO rules and that avoids derailing progress toward free trade and climate goals alike.

Key findings and recommendations

Future national-level policies to address climate change are likely to include provisions that target carbon-intensive products from countries deemed to be making inadequate efforts. These provisions need not violate sensible trade principles and WTO rules, but there is a large danger that in practice they will. The kinds of provisions that would be more likely to conflict with WTO rules and provide cover for protectionism include the following:

- *Unilateral measures applied by countries that are not themselves participating in the Kyoto Protocol or its successors.*
- *Judgments as to findings of fact that are made by politicians, vulnerable to political pressures.*
- *Unilateral measures that seek to sanction an entire country, rather than targeting narrowly defined energy-intensive sectors.*
- *Import barriers against products that are further removed from the carbon-intensive activity, such as firms that use inputs that are produced in an energy-intensive process.*
- *Subsidies – whether in the form of money or extra permit allocations – to domestic sectors that are considered to have been put at a competitive disadvantage, such as those provided for in 2008 by the European Union.*

By contrast, border measures that are more likely to be WTO-compatible include either tariffs or (equivalently) a requirement for importers to surrender tradable permits aligned with the following guidelines:

- *Measures should follow a multilaterally-agreed set of guidelines among countries participating in the emission targets of the Kyoto Protocol and/or its successors.*
- *Judgments as to findings of fact – which countries are complying or not, what industries are involved and what are their carbon contents, what countries are entitled to respond with border measures, or the nature of the response – should be made by independent panels of experts.*
- *Measures should only be applied by countries that are reducing their emissions in line with the Kyoto Protocol and/or its successors, against countries that are not doing so, either as a result of the latter's refusal to join or their failure to comply.*
- *Import penalties should target fossil fuels and a half-dozen of the most energy-intensive major industries: aluminum, cement, steel, paper, glass, and perhaps iron and chemicals.*

Conclusion

A multilateral regime is needed to guide the development of trade measures intended to address concerns about leakage and competitiveness in a world where nations have different levels of commitment to GHG mitigation. Ideally, such a regime would be negotiated along with a Kyoto successor that sets emission-reduction targets for future periods and brings the United States and major developing countries inside. But if that process takes too long, it might be useful in the shorter run for the United States to enter into negotiations with the European Union to harmonize guidelines for border penalties, ideally in informal association with the secretariats of the United Nations Framework Convention on Climate Change and the WTO.

Author affiliation

Jeffrey Frankel is the James W. Harpel Professor of Capital Formation and Growth at the Harvard Kennedy School.

Appendix 1.16 A proposal for the design of the successor to the Kyoto Protocol

Larry Karp and Jinhua Zhao

Overview

This paper proposes a design for a post-2012 international climate agreement (Kyoto II) to follow the Kyoto Protocol. The proposed design would impose national limits on rich countries' greenhouse gas emissions and promote voluntary abatement by developing countries. It includes two new features aimed at promoting participation and compliance and addressing concerns about carbon leakage: (1) an escape clause that would give signatories the option to reduce their abatement requirements in exchange for a penalty and (2) the use of trade restrictions.

Discussion

The primary objective in designing a successor agreement to the Kyoto Protocol is to promote participation and compliance. Other design details are irrelevant if nations do not sign the treaty, or if they sign it but do not honor their commitments. Given limited ability to control the behavior of sovereign nations, an agreement must be designed so that it is in individual countries' interest to participate and comply. A new agreement should also set the stage for future increases in commitment and developing-country participation.

The proposed design for Kyoto II shares many features of the current Kyoto Protocol: it would impose mandatory emissions ceilings on rich countries and preserve existing mechanisms for promoting abatement efforts in developing countries and economies in transition, including Joint Implementation (JI) and the Clean Development Mechanism (CDM). It would also seek to develop sectoral agreements to scale up developing country participation. Nations would be free to choose whatever combination of domestic policies (cap-and-trade, carbon taxes, JI, and CDM, etc.) they prefer to achieve

118

compliance. Developing countries would not face mandatory abatement commitments for the relatively short (at most ten year) period covered by Kyoto II. But to obtain the benefits of participation, they would need to accept the concept of future obligations. The summary below lists other key elements of the proposed design – including new mechanisms to address cost and leakage concerns.

Key findings and recommendations

- *Mandatory emission ceilings:* Rich countries would accept mandatory ceilings under the new agreement. Developing countries would not be subject to abatement requirements, but would be put on notice that they will face obligations in the next round of negotiations. The treaty would encourage voluntary steps and agreements among parties outside Kyoto II while recognizing that these are not substitutes for a multinational agreement with mandatory reduction commitments.
- *Short duration:* A Kyoto II agreement should cover eight to ten years, while stating the goals for future treaties. A relatively short-lived agreement helps account for uncertainty and new information and makes it easier to incorporate changing responsibilities between developed and developing countries.
- *Escape clause:* Nations with mandatory emissions ceilings would have the option to reduce their abatement commitment in exchange for either paying a monetary fine or accepting trade sanctions imposed by other signatories. This mechanism provides protection against unexpectedly high abatement costs, it encourages participation, and it creates incentives for compliance. The severity of the penalty (whether in the form of a fine or trade sanctions) would increase as more nations join, thereby further leveraging incentives for membership and compliance.
- *Trade measures:* In addition to the potential use of trade sanctions in the context of an escape clause, Kyoto II should recognize the use of trade policies in achieving climate-related objectives. Specifically, the authors support the use of carefully circumscribed border tax adjustments to protect against the possibility that firms will re-locate to nations without mandatory commitments to escape carbon costs, thereby undermining the environmental objective and raising competitiveness concerns in signatory nations. This policy can increase incentives for countries to participate instead of staying outside the agreement.

- *Tradable emissions:* Kyoto II should allow international trade in emissions permits. While acknowledging that such trade could marginally reduce incentives to join or levels of abatement undertaken, the authors conclude that the efficiency gains achievable through trade, when countries have different abatement costs, outweigh the potential for perverse effects. Permit trading or allocation should not play an important role in creating inducements for participation. Instead, JI, CDM, and future sectoral agreements should be the primary mechanisms to encourage membership and capture low-cost abatement opportunities.

Conclusion

This proposal builds on the existing Kyoto Protocol but introduces two major new features to address problems of participation and compliance. The inclusion of an escape clause will increase nations' incentive to join, help solve the enforcement problem, and put a ceiling on costs. Some trade economists may dislike the use of trade policy to address leakage concerns and strengthen incentives for participation, but these policies are necessary to overcome objections to the adoption of an effective regime. Negotiating a Kyoto successor with these features would represent an important next step toward the larger objective of bringing all nations under the discipline of a meaningful international climate agreement.

Author affiliation

Larry Karp is Professor, Department of Agricultural and Resource Economics, University of California at Berkeley.
Jinhua Zhao is Associate Professor, Department of Economics and the Department of Agriculture, Food, and Resource Economics, Michigan State University.

Section V

Economic Development, Adaptation, and Deforestation

Appendix 1.17 Reconciling human development and climate protection: a multi-stage hybrid climate policy architecture

Jing Cao

Overview

This paper proposes a multi-stage hybrid climate change policy architecture for the post-2012 era. To reconcile human development and climate protection, a modified Greenhouse Development Right (GDR) burden-sharing formula is proposed and several key design issues are discussed. Finally, we use China as a case study to highlight the role of developing countries in this framework and the challenges to and opportunities for a low-carbon economy.

Discussion

The Kyoto Protocol is an ineffective tool for stabilizing long-term greenhouse gas concentrations. This paper proposes an alternative international climate policy architecture for the post-2012 period. The architecture has three important characteristics. First, the architecture uses a multi-stage framework in which countries face targets and timetables for emissions reductions that vary depending on national circumstances. A multi-stage framework addresses the challenge of long-term target-setting by setting accessible and relatively short-term targets for each stage. It also allows for "common but differentiated responsibilities" among nations – or in other words, allows developing nations to take on emission reduction responsibilities gradually, along with incentives to graduate faster.

The second important characteristic of the architecture is that emissions control responsibility is allocated using a modified "Greenhouse Development Right" burden-sharing formula. This formula defines an individual person's responsibility for reducing emissions as a Cobb-Douglas function of responsibility for cumulative emissions and of capacity to reduce emissions without sacrificing basic necessities. The formula includes two weighting factors that represent ethical judgments about the relative importance of responsibility and capacity. National emission reduction responsibilities are then calculated as the sum of the responsibilities of the citizens of that country. The advantage of the Greenhouse Development Right formula is that it obliges people with incomes and emissions above the threshold to pay the costs of mitigation, while allowing people with incomes and emissions below the threshold to maintain their right to development.

The final important characteristic of the architecture is that its implementation would be negotiated, managed, and enforced using a combination of hybrid top-down and bottom-up organizational structure for diplomatic negotiation. At the global level, an international institution would determine long-term emissions targets. Such an institution could provide a simple negotiation forum that could focus on key emitters and important groups of countries. At the regional level, "clubs" of nations would collaborate and establish regional climate forums for technology transfer, R&D funds, market-based instruments, and enforcement regimes. At the country level, national climate agencies would establish regulatory mechanisms to achieve negotiated obligations.

Key findings and recommendations

A future climate architecture should include the following stages:
- *In the first stage, all member countries would agree on a path of future global emissions that leads to an acceptable long-term stabilization goal.* Developed countries would increase the stringency of their binding Kyoto emissions reduction commitments, and broaden the coverage by including some high-income non-Annex I countries, but developing countries would not be required to make any commitments.
- *In the second stage, developing countries would focus on "no regrets" mitigation options with priorities in local sustainable development.* Sustainable development measures should include gradual phase-out of inefficient and energy-intensive equipment, "no regrets" greenhouse gas mitigation options, and new investments and standards aimed at both development and environmental objectives. At this stage, quantitative targets are still not required, but voluntary emission reductions are strongly encouraged with moderate financial and technological flow from developed countries.

- *In the third stage, developing countries would take on moderate emissions targets that are only binding in one direction.* If these countries achieved their emissions targets, they could sell their allowances, but if they did not achieve their targets, they would face no penalties. This strategy would encourage participation.
- *In the final stage, all countries would agree to binding, absolute emissions targets.* The targets are binding in both directions, thus failure to achieve stated targets will incur stringently enforced penalties.

Such a multi-stage approach can break the current political impasse and start to build trust between the North and South. In addition, an incentive mechanism should be implicitly embodied in this framework. If developing countries choose to graduate faster from one stage to the other, more financial and technological transfer should be provided from the developed countries.

Regarding the GDR burden-sharing formula, this paper improves previous authors' calculations by incorporating cumulative historical carbon emissions back to the 19th century and by taking into account carbon sinks. The results show that:

- *Under the modified GDR burden-sharing calculations, high-income countries should accept 87% of the burden of current emissions reduction targets, middle-income countries should take 13% of the burden, and low-income countries should take 0.1% of the burden.* The United States alone should accept 39% of the burden of current emission targets because of its ability to reduce emissions and its large historical responsibility for emissions. The next largest shares of the burden fall on Germany, Japan, and the United Kingdom, which should accept 7.8%, 7.8%, and 6.4%, respectively. EU (27) as a whole should accept 36% of the burden. The calculations indicate that China should accept 2.2% of the burden, South Africa should accept 0.7%, and India and Brazil have no obligations. Over time, with income increasing in developing countries, more people will exceed the poverty and emission threshold, and thus developing countries will gradually bear more obligations in the future.

Conclusion

Climate change is a complicated problem that requires a fair and effective international climate policy regime. The sooner the current political climate impasse can be broken, the better the future chances of stabilizing the global climate.

Author affiliation

Jing Cao is Assistant Professor in the Department of Economics, School of Economics and Management, Tsinghua University, Beijing, China.

Appendix 1.18   What do we expect from an international climate agreement? A perspective from a low-income country

E. Somanathan

Overview

Although an effective solution to the climate change problem will require the cooperation of developing countries, it is not clear that near-term greenhouse gas emission quotas for these countries are either feasible or desirable. This paper argues that a post-2012 international climate agreement should instead focus on creating incentives to stimulate research and development of new climate-friendly technologies.

Discussion

To limit the rise in global temperature to one or two degrees Celsius will require massive cuts in greenhouse gas emissions by the middle of the century. Since low- and middle-income countries, including China and India, will soon account for more than half of global CO_2 emissions, tackling the climate problem requires that they have the incentive to reduce their emissions substantially. However, since climate change is not a domestic political issue in developing countries, their governments are less likely to cut emissions substantially. Thus, an international agreement cannot realistically demand emissions quotas from developing countries that are below business-as-usual emissions.

This paper argues that an international climate agreement involving developing countries is of secondary importance. A more important priority is generating technological change that lowers the price of alternatives to fossil fuel. There are two ways to stimulate this technological change. First, governments could adopt policies that raise the expected returns to investment in research and development (R&D). These policies could include emissions taxes, cap-and-trade systems, or traditional regulation. Because these policies would force firms to pay for their greenhouse gas emissions, they would create demand for low-carbon technologies and provide financial incentives for firms to invest in R&D. Unfortunately, there is currently little public support for these types of price signals, and so it is difficult to argue that raising energy prices through taxes or cap and trade should be the main instrument of public policy. Furthermore, private investments in R&D are likely to be larger if future demand is more certain. However, relying solely on price incentives in the face of opposing public opinion is not a realistic way of inducing certainty.

A better way to stimulate technological change is to increase public-sector R&D and subsidize private-sector R&D. Global public energy R&D investments have halved in real terms since 1980. However, increasing funding for energy R&D is likely to receive public support simply because it is a mechanism the public understands. For example, although most Americans dislike the idea of higher gas taxes, a majority do support raising taxes if the revenues are used for R&D for new non-polluting energy sources.

Key findings and recommendations

- *It is not clear that emissions trade between developed and developing countries is currently feasible or desirable.* Many developing countries with corrupt governments and weak institutional capacity are unlikely to be able to administer credible domestic cap-and-trade systems. Furthermore, even if some developing countries take on quotas in order sell their emissions permits to richer countries, the resulting increases in energy prices could impact the poor severely. Such price changes could also have perverse effects on climate and human health by encouraging the use of solid fuels for cooking.
- *R&D is a realistic way of making it economical for all countries, including developing countries, to reduce their CO_2 emissions.* R&D has the advantage that resources are spent now, rather than being promised in the future. It also accomplishes a transfer from rich to developing countries to induce the latter to cut emissions, in a manner that is likely to be acceptable to the public in the developed world.
- *A future international climate agreement should take advantage of areas in which there is public support for domestic action.* In developed countries,

this means cap and trade, a greatly increased financial commitment to R&D, and expansion of existing labeling and standards to promote energy efficiency.

- *In the coming round of negotiations, an international agreement involving developing countries should confine itself to promoting technical cooperation.* The bulk of the finance for developing country action will have to come from developed countries. Energy conservation and agriculture research are areas in which developing countries would see significant economic co-benefits, so these are areas that should be included in an agreement. In many developing countries there is also considerable interest in improvements in energy efficiency. A formal agreement for sharing expertise and information between regulatory agencies would also improve the quality of many countries' regulation of energy, greenhouse gases, and associated pollutants. In addition, an agreement should support a major thrust in tropical agriculture to develop new varieties that will withstand climate change.

Conclusion

The major action that is needed to realize huge emissions reductions over the next few decades is the promotion of R&D that will make carbon-neutral energy sources much more competitive with fossil fuels. Developed countries will have to promote this R&D not only through domestic regulation, taxes, and tradable permits, but also by committing more government funds to R&D for non-carbon energy sources.

Author affiliation

E. Somanathan is Professor in the Planning Unit, Indian Statistical Institute, in New Delhi, India.

Appendix 1.19 Climate accession deals: new strategies for taming growth of greenhouse gases in developing countries

David G. Victor

Overview

Managing the dangers of global climate change will require developing countries to participate in a global climate regime. So far, however, those nations have been nearly universal in their refusal to make commitments to reduce growth in their greenhouse gas emissions. This paper describes how a set of international "Climate Accession Deals" could encourage large policy shifts in line with the interests of developing countries and, furthermore, how they could reduce greenhouse gas emissions.

Discussion

Conventional "carrot and stick" strategies for engaging developing countries are unlikely to be effective. The biggest carrot – the Kyoto Protocol's Clean Development Mechanism (CDM) – requires difficult judgments on whether emissions reductions are additional, is plagued by high transaction costs, and creates perverse incentives for developing countries to avoid policies that would reduce emissions. Threats of climate-related trade sanctions could, in theory, encourage developing countries to do more. But such tools are blunted by their questionable legal basis, high administrative costs, and adverse impacts on the world trading system. Although these approaches could be fine-tuned, a fundamentally different approach will be needed.

This paper suggests a new strategy for engaging developing countries, focusing on the concept of Climate Accession Deals (CADs). These deals would take advantage of the fact that there are many large policy shifts that are in these countries' interests and which also, fortuitously, reduce greenhouse gases. Each CAD would include a set of policies that are tailored to gain maximum leverage on a single developing country's emissions. In addition, the policies included in each CAD would align with a country's interests and capabilities so that the initial investments are easily expanded with few incentives for developing countries to abandon the effort once under way. Industrialized countries would support each CAD by providing specific benefits such as financial resources, technology, administrative training, or security guarantees.

The closest analogy to CADs is the accession process to the World Trade Organization (WTO). As with WTO accession agreements, CADs would be complex to engineer (thus must be few in number and must initially focus on the countries with the highest potential for reducing emissions) and would not be an end point to engagement with developing countries. Rather, they would frame a long-term transition during which the developing country would become a full member of the climate regime – eventually setting targets that span the entire economy and adopting complementary policies.

Key findings and recommendations

- *Compared to conventional approaches, Climate Accession Deals have several important advantages.* First, they are anchored in host countries' interests and capabilities and thus do not require negotiating agreements that run contrary to a country's interests. Second, they are limited in number and can be crafted through a bidding process that would allow maximum leverage while minimizing external investment. Third, all involve a complex array of interests and institutions, and thus would engage more than the environmental and foreign affairs ministries that have dominated climate diplomacy to date. Fourth, all CADs are replicable and scalable. Where they succeed, they offer paths for similar deals (at lower cost) in other countries, and are self-reinforcing in the original host country.

  Examples of potential Climate Accession Deals include:
  - *A CAD for China should focus on carbon emissions from coal,* which accounts for 69 percent of its primary energy system. Recent energy shortages and higher energy prices have caused Chinese officials to initiate programs to decouple economic growth from energy consumption. A CAD could encourage this goal by making new power plants more efficient, encouraging greater use of natural gas and nuclear fuel, improving

the efficiency of the electric power grid, and funding research projects on new systems for electric supply. Improved efficiency of the power supply system could save more than 200 million tons of CO_2 annually.

o *A CAD for India should also seek ways to use coal more efficiently and supplant coal.* The greatest opportunity for leverage on India's emissions lies in boosting the efficiency of converting coal to electricity. Other energy sources, including hydro, wind, gas, and nuclear power, could also make a difference at the margin. Improving the country's weak system for power delivery would reduce greenhouse gas emissions, as would expanding access to electricity to reduce particulate emissions from traditional biomass usage.

o *In South Africa, a CAD supporting deployment of advanced power plants* might save 50-100 million tons of CO_2 annually by 2025. A carbon storage scheme might increase that amount another 20 million tons.

o *CADs for Brazil and Indonesia should focus on preventing and reversing deforestation.* One area of assistance could be the provision of surveillance radar, drones, and helicopters for a much larger police force. Such systems would allow these countries to better use existing personnel to monitor deforestation and regulate illegal logging.

o *A CAD for Russia should target flaring of natural gas.* Flaring at Russian oil operations releases the equivalent of 175 million tons of CO_2 annually.

Conclusion

A serious strategy for engaging developing countries in a global climate agreement must contend with two truths. First, at present, developing nations value economic growth far more than future global environmental conditions. Second, many governments of developing nations have little administrative ability to control emissions. Climate Accession Deals could address both of these barriers to effective engagement of developing countries in a way that is consistent with those countries' incentives and effective in putting them on a path to fuller commitments to reduce emissions.

Author affiliation

David G. Victor is Professor of Law at Stanford Law School and Director of the Program on Energy and Sustainable Development.

Daniel S. Hall, Michael A. Levi, William A. Pizer, and
Takahiro Ueno

Overview

A global effort to mitigate climate change will require cooperation between
developed and developing countries. Even as many developed countries are
at some stage of enacting significant domestic regulations to meet global
stabilization goals, growth in developing country emissions will easily thwart
those goals unless a cooperative solution is found. This paper argues that
a wide range of options should be pursued to increase developing-country
mitigation efforts over time.

Discussion

Much of the current post-Kyoto policy debate focuses on what role developing countries should play vis-à-vis industrialized countries in terms of reducing global emissions, while achieving sustainable economic growth, and how
wealthy nations can best support and encourage mitigation efforts by poor
nations. Because climate change is fundamentally a global problem, the participation of all major emitting countries is essential, and developing countries are an important – indeed the most important – source of emissions
growth over the next century.

Developing countries typically place greater priority on economic development than on environmental protection, despite being vulnerable to the
potential adverse effects of continued warming. Countries like the United
States, meanwhile, understand that their own emissions mitigation efforts
can be negated if, through open trade in goods and services, their emitting
activities shift to non-participants. This would not only undermine the environmental objective, it would also raise serious jobs and competitiveness
concerns.

Key findings and recommendations

This paper identifies three categories of options for increased developing country engagement: domestic policy reforms in developing countries, expanded financing mechanisms to address incremental costs, and diplomatic efforts.

- *Domestic policy reforms can produce direct economic, political, and environmental benefits in developing countries while simultaneously contributing toward GHG mitigation. "Win-win" policy opportunities often exist with respect to energy subsidies, energy efficiency, and technology transfer.* For example, fossil fuel subsidies are common in many developing countries. Reducing or eliminating them would relieve budget pressures, promote more efficient energy use, improve energy security, and avoid unintended distributional consequences while also slowing the growth of GHG emissions. Similarly, energy efficiency improvements provide multiple benefits and can be encouraged through a variety of mechanisms. Finally, governments can use information, regulation, pricing strategies, investment, and a variety of other levers to promote the more rapid and thorough diffusion of climate-friendly technologies. For their part, developed countries can support climate-friendly domestic policy reforms in developing countries by providing technical assistance and financial support, partnering with relevant agencies, and building human and institutional capacity.

- *The scale of investment needed in non-OECD countries to achieve global GHG mitigation objectives necessitates a rethinking of the international financing mechanisms available for transferring resources from developed to developing countries.*

  o *Offset mechanisms* have the advantage that they channel private resources to support mitigation efforts in developing countries. The largest existing program of this type is the Clean Development Mechanism (CDM). It has been criticized for providing excessive subsidies, supporting a limited number of project types, and channeling most funds to large developing countries that may have less need of assistance than poorer countries. Two categories of CDM reforms are currently being discussed: (1) adjustments to crediting rules to address the distribution of projects and subsidy levels and (2) moving beyond project-based credits to provide credits for programs, policy reforms, or sectoral targets.

  o *Public funds* have the advantage that they can provide financial support up front (as opposed to only after a project is generating reductions). They also give funders greater control over the amount of subsidy provided to different types of mitigation activities and greater flexibility in terms of how the support is provided (e.g., grants vs. loans vs. loan guarantees). On the other hand, this financing mechanism is subject to budget appropriations by donor governments. Governance is also often contentious, with donors and recipients vying for greater control over the use of international public funds.

- *Both of the options already discussed, domestic policy reforms and international financing, are inextricably tied to public and private diplomacy.* Financing and other forms of assistance are an important source of leverage in diplomatic efforts. Others include energy security, the threat of trade sanctions, support for adaptation efforts, and broader forms of linkage (for example, Russia ratified the Kyoto Protocol after EU countries agreed to ease Russia's passage into the World Trade Organization). Coordination and cooperation among key institutions is also critical to the success of diplomatic efforts: the United Nations Framework Convention on Climate Change process remains the primary vehicle for international climate diplomacy, but other multilateral and regional forums are increasingly also playing a role.

Conclusion

Post-Kyoto international climate negotiations are likely to focus on a "grand bargain" with developing countries, offering some form of commitments in exchange for further emission reductions and increased financing from developed countries. Developing country commitments could take the form of domestic policy reforms, sectoral targets, or even economy-wide limits. Because no single approach offers a sure path to success, a variety of strategies – including policy reforms, financing approaches, and diplomatic avenues – must be pursued simultaneously.

Author affiliation

Daniel S. Hall is Research Associate at Resources for the Future.
Michael A. Levi is Senior Fellow at the Council on Foreign Relations.
William A. Pizer is Senior Fellow at Resources for the Future.
Takahiro Ueno is a researcher at the Central Research Institute of the Electric Power Industry, Tokyo, Japan.

Appendix 1.21 International forest carbon sequestration in a post-Kyoto agreement

Andrew J. Plantinga and Kenneth R. Richards

Overview

Forest carbon management must be an important element of any international agreement on climate change. A "national inventory" approach, in which nations receive credits or debits for changes in the national forest carbon inventory, would offset global CO_2 emissions more effectively than the current Kyoto Protocol system.

Discussion

Forest carbon flows comprise a significant part of overall global greenhouse gas emissions. While global forests as a whole may be a net carbon sink, global emissions from deforestation contribute between 20 and 25 percent of all greenhouse gas emissions. Furthermore, the amount of carbon sequestered in forest vegetation is approximately 359 billion tons, compared to annual industrial carbon emissions of 6.3 billion tons. Thus, natural and anthropogenic changes to forests could have an enormous impact on the global carbon cycle.

An effective international forest carbon management system must provide landowners and governments incentives to protect and expand forest stocks of carbon. The Kyoto Protocol has proven ineffective in this regard. For one, the Kyoto approach discourages countries from accepting the responsibility for all carbon changes under their authority. Second, the Protocol's Clean Development Mechanism (CDM) fails to provide developing countries with incentives to reduce carbon emissions through forest management. Since the carbon effects of individual forestry projects are difficult to measure, the CDM is a poor tool to provide incentives for forestry policy. Finally, the Kyoto Protocol approach may actually accelerate deforestation by shifting timber harvesting from Annex I to non-Annex I countries.

The expiration of the Kyoto Protocol in 2012 invites a reexamination of how to address terrestrial carbon management within the framework of an international climate change treaty. There are three basic policy approaches.

The first approach, which is currently used by the CDM, is project-level accounting. Individual landowners can apply for credits for net increases in carbon stored in forests on their land. Once the permitting authority verifies the carbon sequestration is valid, the landowner can then sell the credits in permit markets. Unfortunately, past experience has shown that project-by-project accounting faces serious challenges, including the difficulty of establishing the counterfactual reference case against which to evaluate projects.

A second, more promising approach is national inventory accounting. Under this approach, nations conduct periodic inventories of their entire forest carbon stock. The measured stock is compared to a pre-negotiated baseline stock to determine the offset credits that can be redeemed, or debits that must be covered, in the linked emissions allowance market. To avoid the difficult task of forecasting future stocks, international negotiations determine the reference stock. This process allows countries to address equity and fairness issues and to provide incentives for participation in the agreement.

A final policy approach is one in which forest carbon programs are not linked to the emissions allowance system. Rather than focusing on carbon credits, the program would focus on inputs such as policies to discourage deforestation, programs to encourage conversion of marginal agricultural land to forests, and projects to better manage forests in forest-rich countries. These commitments would be incorporated into the national plans required under the UNFCCC and would be funded by overseas development aid, international institutions, or through a separate climate fund. Such a delinked system would have several advantages, including lowering transaction costs, but it would also have two particularly serious disadvantages: dulled incentives for landowners and the loss of private sector funding.

Key findings and recommendations

- *The Kyoto Protocol's current forestry policy – project-level accounting – should not be a central component of the forestry mechanisms adopted in a post-2012 agreement.* This approach has fundamental flaws, including problems with additionality, permanence, leakage, and adverse selection.
- *The national inventory approach greatly reduces the problems that plague the CDM's project-level accounting.* It also provides comprehensive coverage

of changes to forest carbon stocks and can be used to address equity and fairness concerns.

- *The national inventory approach also has disadvantages that need to be acknowledged.* First, the scope of carbon sequestration activities is limited to those that will show up in a national forest carbon inventory. Second, the approach provides incentives for governments, not private project developers, which may be a disadvantage in countries with weak governmental institutions, corruption, or powerful special interest groups. Third, problems with additionality, permanence, etc. may resurface with domestic carbon sequestration policies pursued by national governments.
- *A delinked, input-based approach could be used as an interim measure if current measurement technologies are inadequate to support the national inventory approach.* The national inventory approach requires regular and reliable national forest inventories for a large group of countries. An input-based approach could be used temporarily, while the scientific community works to further develop the measurement capacity necessary to support national inventories.

Conclusion

The national inventory approach is an attractive alternative to the Kyoto Protocol's existing forestry approach.

Author affiliation

Andrew J. Plantinga is Professor of Agricultural and Resource Economics at Oregon State University.

Kenneth R. Richards is Associate Professor of Environmental Science at Indiana University's School of Public and Environmental Affairs.

Section VI

Modeling Impacts of Alternative Allocations of Responsibility

Appendix 1.22 Modeling economic impacts of alternative international climate policy architectures: a quantitative and comparative assessment of architectures for agreement

Valentina Bosetti, Carlo Carraro, Alessandra Sgobbi, and
Massimo Tavoni

Overview

With broad recognition that a coordinated global effort is needed to address climate change, negotiations are already under way to define a new international climate agreement. Various architectures for such an agreement have been proposed. This paper undertakes a first-of-its-kind comparison of some prominent options using a common framework to assess four features of these architectures: economic efficiency; environmental effectiveness; distributional implications; and political acceptability, as measured in terms of feasibility and enforceability. The aim is to derive useful policy insights for designing a post-Kyoto agreement.

Discussion

The policy architectures compared in this paper are summarized below. Each makes different trade-offs in scope and timing, with implications for cost, environmental effectiveness, and political feasibility.

1. Global Cap and Trade with Redistribution: In this benchmark scenario, all nations participate immediately in a global cap-and-trade system designed to stabilize atmospheric CO_2 at 450 parts per million (ppm) by 2100. Permits are allocated to all countries on an equal per capita basis.

2. Global Tax Recycled Domestically: All countries apply a globally-consistent carbon tax designed to achieve the same stabilization trajectory as above. Revenues from the tax are recycled domestically and implementation begins immediately.

3. Reducing Emissions from Deforestation and Degradation: Same as the first scenario, except credits from avoided Amazon deforestation are included in the permit market.

4. Climate Clubs: In this scenario, a group of mostly advanced economies agrees to abide by its Kyoto target and reduce GHG emissions 70 percent below 1990 levels by 2050. Other fast-growing countries and regions begin gradual efforts to reduce emissions below business-as-usual (BAU), but converge to the same level of reductions as the first group after 2050. All remaining countries face no binding targets, but their emissions are limited to BAU.

5. Burden Sharing: Developed (Annex 1) countries commence abatement immediately, with the burden shared on an equal per capita basis. Binding emissions targets are extended to all other countries, except those in sub-Saharan Africa, in 2040.

6. Graduation: Countries adopt binding emission targets as they reach specified criteria for income and emissions. Annex 1 countries compensate for the delayed entry of non-Annex 1 countries by undertaking additional reductions as required to achieve a 450 ppm stabilization trajectory.

7. Dynamic Targets: Different countries adopt different targets over time depending on current and projected emissions, income, and population.

8. R&D and Technology Development: No binding emissions targets; instead, all countries contribute a fixed percentage of GDP to an international fund for developing low-carbon technologies.

Key findings and recommendations

The authors apply the WITCH climate-energy-economy model to assess each proposed architecture along four metrics: environmental effectiveness (measured as expected temperature change above pre-industrial levels in 2100); economic efficiency (measured as change in gross world product, or GWP, relative to BAU); distributional implications (using the Gini Index to measure income inequality across different regions of the world in 2100); and potential enforceability (measured by changes in global and regional welfare with respect to the status quo).

Several policy-relevant insights emerge:

- *All the policy architectures evaluated in this analysis produce warming above the 2°C target envisaged by the IPCC and the European Commission.* More drastic measures than any of those modeled for this analysis will be required to meet the target.
- *There is a clear trade-off between environmental effectiveness and cost.* The inclusion of credits for avoided deforestation helps reduce cost somewhat, but estimated gross world product (GWP) losses in all the scenarios designed to achieve CO_2 stabilization at 450 ppm (Scenarios 1, 2, 3, 5, 6) exceed 1 percent. The Climate Clubs and Dynamic Targets scenarios (4, 7) are significantly less costly, but also less effective. The R&D-only scenario (8) actually leads to slight gains in GWP, but it is also the least effective in terms of reducing emissions.
- *There is a clear trade-off between environmental effectiveness and enforceability.* If one assumes that countries' willingness to participate will depend on the expected welfare effects of the policy, then the more stringent architectures will also be the most difficult to enforce.
- *Any of these architectures would produce a more fair distribution of income in 2100 relative to the current situation.* In the more stringent scenarios (i.e., those designed to stabilize CO_2 at 450 ppm), however, these gains in equality occur in the context of significant overall GDP losses. Of the architectures modeled, the most egalitarian are Climate Clubs, Graduation, and Dynamic Targets (4, 6, 7) because they distribute the abatement burden according to per capita income and emissions. The inclusion of credits for avoided deforestation (3) also improves equity because most forest-related abatement opportunities are located in developing countries.

Conclusion

From a cost and enforceability standpoint, GHG stabilization at 450 ppm for CO_2 only (550 ppm CO_2-equivalent for all GHGs) is hardly achievable. However, there may exist a strategy of progressive commitments, in which future binding targets are set to achieve consensus in developing countries, whereas developed countries move first, that can achieve GHG stabilization very close to 450 ppm. An extended – possibly global – carbon market, even without global commitments to reduce emissions, helps reduce costs. In addition, including non-CO_2 gases and credits for avoided deforestation further reduces costs. Nonetheless, a basic trade-off between economic impact and environmental protection remains.

Author affiliation

Valentina Bosetti is a Senior Researcher and Massimo Tavoni a Senior Researcher at Fondazione Eni Enrico Mattei (FEEM), Venice, Italy; both are also affiliated with the Princeton Environmental Institute at Princeton University.

Carlo Carraro is Professor at the University of Venice and Research Director of FEEM.

Alessandra Sgobbi is a Researcher at FEEM.

Appendix 1.23 Sharing the burden of GHG reductions

Henry D. Jacoby, Mustafa H. Babiker, Sergey V. Paltsev, and
John M. Reilly

Overview

Success in upcoming climate negotiations will require a clear-eyed view of
the relationship between emissions targets and equity goals. This paper uses
the MIT Emissions Prediction and Policy Analysis (EPPA) model to estimate
the welfare and financial implications of various distributional and emissions
reduction outcomes. The results indicate that a target of reducing global
emissions by 50 percent by 2050 while meeting equity targets is extremely
ambitious, and would require large financial transfers from developed to
developing countries.

Discussion

International climate negotiations have produced a variety of emissions
reduction goals. For example, the G8 has proposed reducing global emissions
by 50 percent by 2050. At the same time, there is a general sense that developed countries should take a disproportionate share of the reduction burden,
with goals of 70 to 80 percent reductions by 2050 expressed by a number
of them, plus there are calls for positive incentives for mitigation action by
developing nations and for protection from the indirect (e.g., terms-of-trade)
effects of mitigation elsewhere.

To assess the compatibility of these environmental and equity goals,
this paper uses the EPPA model to explore the mitigation effort and financial transfers that would be necessary, using a cap-and-trade system as an
example. Burden-sharing agreements are represented by initial allowance
allocations and subsequent emissions trade. One output of the calculation
is the size of the financial transfers that developed countries would have to
make to compensate developing countries. Several burden-sharing rules
are assessed, including allocating the entire burden to developed countries,

allocating burden based on specific equity goals for developing and developed countries, and allocating burden based on simple allocation rules implied by current proposals.

Key findings and recommendations

- *The proposed target of a 50 percent global reduction below 2000 emissions level by 2050 is very ambitious.* Absent near universal participation, such a goal is not achievable, because economic activity and emissions would shift to nations that do not sign the agreement. Even with all nations taking on commitments, this goal would require a complex web of financial transfers to simultaneously satisfy widely-discussed burden-sharing goals.
- *Two interacting equity concerns would have to be dealt with in seeking the global emissions goal.* First, incentives and compensation for developing country participation will be required. Second, since mitigation costs and compensation payments by developing countries will be substantial, they also will need to find an acceptable burden-sharing arrangement among themselves.
- *Simple emissions reduction rules are incapable of dealing with the highly varying circumstances of different countries.* For example, a 30-70 proposal (in which developing and developed nations reduce their emissions by 30 percent and 70 percent, respectively, by 2050) initially appears to be a generous offer from the developed countries. However, it turns out that it could result in net purchases of allowances by some developing countries from the developed countries, in effect partially compensating the richer ones for their mitigation and counteracting the intent of the Bali Action Plan.
- *Other simple rules, such as distributing allowances on a per capita basis or inversely related to GDP per capita, would shift costs toward richer countries but raise other difficulties.* For some developing countries, they produce large net benefits beyond costs related to mitigation. For others, they are even more costly than the terms of the 30-70 proposal.
- *Successful climate negotiations will need to be grounded in a full understanding of the substantial amounts at stake.* If developing countries are fully compensated for the costs of mitigation in the period to 2050, then the average welfare cost to developed countries is around 2 percent of GDP in 2020 (relative to reference level), rising to 10 percent in 2050. The implied financial transfers are large – over $400 billion per year in 2020

and rising to around $3 trillion in 2050. The United States' share of these transfers would be $200 billion in 2020, and over a trillion dollars in 2050.

- *With less than full compensation, the welfare burden on developing countries would rise, but the international financial transfers would remain of unprecedented scale.* It is an extreme assumption that developing countries will demand complete compensation. If, as is likely, they are willing to bear some costs, then the welfare burden on the developing countries will be reduced. Also, the burden is lowered somewhat if compensation only covers direct mitigation costs and not other losses associated with the policy, as might come through terms-of-trade effects. In the process, the required financial transfers are reduced as well, but they remain large.

Conclusion

The G8 countries have called for an aggressive global emissions goal, while previous climate agreements and the Bali Action Plan create an expectation of incentives and compensation for developing country participation. Unfortunately, the magnitude of the implied international financial transfers suggests that this combination of targets may have been chosen without sufficient regard for the difficulty of finding a mutually acceptable way to share the economic burden.

Author affiliation

The authors are all with the Joint Program on the Science and Policy of Global Change at the Massachusetts Institute of Technology (MIT). Henry D. Jacoby is Co-Director and Professor of Management in MIT's Sloan School of Management; Mustafa H. Babiker is Research Associate; Sergey V. Paltsev is Principal Research Scientist; and John M. Reilly is Associate Director for Research and Senior Lecturer in the Sloan School.

Appendix 1.24 When technology and climate policy meet: energy technology in an international policy context

Leon Clarke, Kate Calvin, Jae Edmonds, Page Kyle, and Marshall Wise

Overview

Both the nature of international climate policy architectures and the availability of new energy technologies will influence the costs and regional character of global greenhouse gas emissions reductions. This paper explores the implications of interactions between technology availability, on the one hand, and performance and international policy architectures, on the other, for technology deployment and mitigation costs associated with reaching a long-term atmospheric goal. Key issues explored in the paper include the relative importance of technological improvements with and without efficient international policy, the linkage between policy and technology availability and the long-term evolution of the energy system, the indirect benefits of national R&D efforts through international technology diffusion, overshoot pathways to stabilization, and the role of bioenergy production with CO_2 capture and storage (CCS).

Discussion

Technology and policy are both important to limiting anthropogenic climate change. The development of cheaper and more effective technologies will be critical for reducing the costs and increasing the social and political viability of substantial greenhouse gas emissions reductions. Policy structures influence the overall costs of mitigation along with the regional and sectoral distributions of these costs. This paper analyzes the relationship between international policy architecture and technological development. It simulates the global costs of a variety of policy and technology scenarios, using the MiniCAM integrated assessment model. Four technology development scenarios are explored: (1) a reference scenario under which technological improvement is modest and neither nuclear power nor CCS deploys beyond today's levels, (2) a scenario under which only bioenergy, CCS, and hydrogen technologies all improve rapidly, (3) a scenario under which only renewable, nuclear, and energy efficiency technologies all improve rapidly, and (4) an "advanced" technology scenario under which all of the above technologies improve rapidly.

The analysis combines these technology scenarios with two alternative international policy architectures: one with full participation by all nations from 2012 onward and another with delayed and incomplete participation by developing nations. Both architectures are designed to limit the concentration of atmospheric CO_2 to 500 ppm in the year 2095. By construction, the first is more economically efficient, but unlikely to be realized. The second is less efficient and shifts emissions mitigation toward presently developed regions and away from developing regions.

Key findings and recommendations

- *Technology is even more important to reducing the costs of emissions mitigation when international policy structures deviate from immediate and full participation.* In simulations, the global cost savings from advanced technology are twice as large when participation in the international agreement is incomplete. Given that real international policy architectures will undoubtedly deviate from full efficiency, this result further reinforces the need for technology policy as a foundational component of national and international mitigation policy.

- *International diffusion of climate technology may be as or more important to domestic mitigation cost containment as domestic technology diffusion.* The vast majority of technology research is conducted at the national level. However, development and diffusion of climate change technology is a global public good, since it reduces future global emissions. This implies that there is a strong, albeit indirect, incentive for individual nations to implement structures that would enhance international technology diffusion in order to help achieve long-term environmental goals with the least degree of national mitigation. The simulations demonstrate that these indirect benefits may, depending on burden-sharing and other elements of the international architecture, be more important than the direct benefits associated with reducing costs to meet a particular national emissions trajectory.

- *Given a particular long-term climate goal, policy architecture has a larger impact on the distribution of mitigation actions than on the global emissions pathway.* For the mitigation level considered in this paper, there is limited flexibility in the global emissions pathway relative to the flexibility in the distribution of emissions among regions. Delays in participation by developing regions imply roughly commensurate increases in mitigation activities, along with increases in developed region carbon prices and the deployment of low-emissions technologies in these regions.

- *More rapid technology improvements reduce the relative influence of policy architecture.* Additionally, although regional emissions are more sensitive to international policy architecture, technology availability remains a strong force shaping emissions, regardless of the international policy architecture. Indeed, some degree of emissions mitigation in developing countries may be viable in the near term without explicit climate policies in those countries, simply by easing the deployment of technologies with economic or other benefits that accrue absent climate policy, such as advanced end-use technologies, nuclear energy, and renewable power.

- *Combining the use of bioenergy with CCS opens the possibility of electricity production with net negative global carbon emissions.* The combination of bioenergy and CCS allows for negative emissions, easing mitigation costs. As carbon prices rise, bioenergy is increasingly deployed in power generation with CCS, eclipsing the use of bioenergy as a liquid fuel and eventually becoming the dominant use of bioenergy.

Conclusion

Limiting CO_2 concentrations will require movement to a very different future energy system than that of today. Uncertainty in future international policy architectures provides further justification for the inclusion of technology instruments as a prominent element of national and international climate policy. The focus of the near term should be on preparation for dramatic transformations of the energy system through technology experimentation, exploration, and development. Near-term technology-related actions to take advantage of energy technology's potential are (1) to begin to reduce emissions through technology deployment, (2) to make investments in science, technology, and human capital resources to maximize the number of long-term options for mitigation, and (3) to ascertain which will be the most effective long-term options and guide the technology portfolio accordingly.

Author affiliation

All of the authors are at the Pacific Northwest National Laboratory's Joint Global Change Research Institute:
Leon Clarke, Senior Research Economist
Kate Calvin, Research Economist
Jae Edmonds, Chief Scientist and Laboratory Fellow
Page Kyle, Research Analyst
Marshall Wise, Senior Research Scientist

Appendix 1.25 Revised emissions projections for China: why post-Kyoto climate policy must look East

Geoffrey J. Blanford, Richard G. Richels, and
Thomas F. Rutherford

Overview

Carbon dioxide (CO_2) emissions from China's energy sector have recently grown much faster than most analysts expected. This paper presents updated global emissions projections after taking into account recent trends in China and other developing countries. The results suggest that, even allowing for a near-term global recession, unconstrained emissions growth in the developing world will very rapidly put global stabilization targets in jeopardy.

Discussion

Since 2000, China's energy-related CO_2 emissions have grown at an average rate of more than 10 percent per year. Global forecasts have been slow to reflect this trend, in part because it follows two decades in which the energy intensity of the Chinese economy declined with privatization and the introduction of market reforms. But rapid growth in industrial demand and a heavy reliance on coal have since combined to reverse that decline. In 2006, China surpassed the United States as the world leader in carbon emissions.

More broadly, it now appears that emissions from developing countries as a group will exceed those of developed countries before 2010, much earlier than previously expected. To explore the implications of these trends, the authors recalibrate MERGE, a general equilibrium model of the economy and energy use, to incorporate more recent data from China. Comparing the results to historic emissions and development patterns in similar Asian countries, the authors find that their new projections for China are largely consistent with past experience.

Higher baseline or "business-as-usual" (BAU) emissions make any given target for stabilizing atmospheric CO_2 concentrations more difficult and expensive to achieve. Comparing their updated emissions projections to different stabilization pathways and considering three distinct scenarios for developing-country participation in future mitigation efforts, the authors reach sobering conclusions about the attainability of commonly discussed CO_2 stabilization targets.

Key findings and recommendations

- *Updated projections for China suggest that continued growth in developing-country emissions could put stabilization targets effectively out of reach within the next 10 to 20 years, regardless of what wealthier countries do.* By as early as 2020 or 2030, depending on the stabilization target chosen, BAU emissions from developing countries alone – even with zero contribution from developed (so-called "Annex B") countries – could exceed the maximum consistent with a reasonable trajectory for achieving CO_2 stabilization.
- *A CO_2 stabilization target of 450 parts per million by volume (ppmv) is probably only possible with immediate action in all countries, while a target of 550 ppmv now appears almost as challenging as 450 ppmv appeared just a few years ago.* To achieve the 450 ppmv target, even with all nations participating under optimal conditions, abatement must begin immediately, such that global emissions by 2020 are 25 percent below BAU levels. This implies a carbon price of $335 per ton carbon in 2020, or $90 per ton CO_2.
- *Stabilization at 550 ppmv may still be feasible* if developed countries undertake immediate reductions and developing countries follow a "graduated accession" scenario in which China and other mid-income countries (e.g., Korea, Brazil, Mexico, and South Africa) join global mitigation efforts in 2020, India joins in 2040, and poorer countries delay participation until 2050. However, this requires that (a) developed countries achieve 50 percent reductions (below 2005 emissions) by 2050 and (b) developing countries, when they join, join at the same level of stringency as developed countries, which translates to adopting a carbon price trajectory that rises from $38 per ton in 2020 to $160 per ton in 2050.
- *If developing countries enter into a global regime more gradually – for example, by adopting progressively more stringent targets only as incomes*

rise – global emissions continue to grow through 2050 and even the 550 ppmv target begins to look doubtful. To explore this option, the authors use a simple correlation between country-level income and targets adopted during the first Kyoto commitment period. This produces substantially better results than the no-participation case, but global emissions are still higher than in the graduated accession scenario and well above the optimal trajectory for stabilization at 550 ppmv.

- *The longer developing countries continue on a BAU path, the more globally costly it becomes to achieve a given stabilization target.* At the same time, participation in a global regime can impose disproportionate costs on developing economies. The authors find that rapidly growing countries like China and India experience a higher percent reduction in GDP for the same percent reduction in emissions under an optimal 550 ppmv stabilization scenario where all countries participate at the same level of stringency. Thus, equity-based adjustments may be needed to motivate developing-country participation in the timeframe needed to achieve stabilization objectives.

Conclusion

The recent acceleration of energy-related emissions in the developing world, particularly China, has taken many analysts by surprise. An updated view of global trends suggests that stabilization at the 450 ppmv level is not an option without full global participation beginning today, while achieving a 550 ppmv target will be much more difficult than previously anticipated, particularly with only partial participation. In this context, it is urgent that international climate negotiations establish incentives for timely and meaningful participation by developing countries, especially China.

Author affiliation

Geoffrey J. Blanford is Program Manager and Richard G. Richels Senior Technical Executive for the Global Climate Change Research Program at the Electric Power Research Institute.

Thomas F. Rutherford is Professor of Economics at the Swiss Federal Institute of Technology in Zurich.

Appendix 1.26 Expecting the unexpected: macroeconomic volatility and climate policy

Warwick J. McKibbin, Adele Morris, and Peter J. Wilcoxen

Overview

Because any effective international climate policy will need to be in place for decades, many unexpected macroeconomic shocks will occur during the policy's existence. This paper explores how such shocks affect global economic conditions and carbon dioxide emissions under two canonical climate policy regimes: a global constraint on the quantity of emissions, such as a cap-and-trade system, and a global policy that equalizes marginal costs of carbon abatement, such as a harmonized carbon tax or a hybrid policy. The results show that a global quantity constraint can have unintended side effects by causing an economic boom in one part of the world to lead to a recession in other areas. A price-based system, on the other hand, tends to exacerbate economic downturns. A hybrid policy can potentially avoid both problems.

Discussion

The global financial crisis, a looming global recession, and turmoil in credit markets drive home the importance of developing a global climate architecture that can withstand major economic disruptions. Such an architecture needs to be resilient to large and unexpected changes in economic growth, technology, energy prices, demographic trends, and other factors that affect the costs of emissions abatement. The stability of global policy has important environmental implications for two reasons. First, collapse of the policy could set back progress on emissions reductions for years. Second, since the decisions of economic actors depend on their expectations of future policy, policies that are vulnerable to shocks are less likely to encourage long-term investments in new technologies and emissions reductions.

This paper uses a computational macroeconomic model to explore how shocks in the global economy propagate under two different climate policy regimes: a cap-and-trade system and a price-based mechanism such as a globally-harmonized carbon tax or hybrid system of national long-term and short-term permits. The paper examines two kinds of shocks relevant to recent experience: (1) a positive shock to economic growth in China, India, and other developing countries, and (2) a sharp decline in housing markets and a rise in global equity risk premiums, causing severe financial distress in the global economy. The paper compares economic results under the two climate regimes in the decade following the shock and draws inferences about which regime offers the strongest incentives to sustain participation in the climate agreement.

The results show that, although the climate regimes appear to be similar in their ability to reduce carbon emissions efficiently, they differ in how they affect the transmission of economic disturbances between economies. These differences have important implications for the political robustness of these policies.

Key findings and recommendations

- *All else equal, a climate regime that exacerbates downward macroeconomic shocks or depresses the benefits of positive macroeconomic shocks would be more costly and less stable than a system that better handles global business cycles and other volatility.* This occurs because macroeconomic shocks can cause the cost of regulation to be much higher or lower than anticipated. These economic surprises can subject governments to enormous pressures to relax or repeal taxes or other policies perceived to impede economic growth.
- *It is critical to get the global and national governance structures right.* There must be a clear regulatory regime in each country and a transparent way to smooth out excessive short-term volatility in prices. A system that enables or even encourages short-term financial speculation in climate markets may collapse at huge expense to national economies. A hybrid system provides many of the advantages of a permit system while limiting opportunities for speculation through the annual permit mechanism.
- *Since shocks in one part of the world will certainly occur, the global system needs to have adequate firewalls between national climate systems to prevent destructive contagion from propagating local problems into a system-wide*

failure. A global cap-and-trade system would be very vulnerable to shocks in any single large economy. A system based on national hybrid policies, on the other hand, would be explicitly designed to partition national climate markets and limit the effects of a collapse in climate policy in one part of the world on climate markets elsewhere.

- *In many respects, a global cap-and-trade system is less robust to macroeconomic shocks than a carbon tax or hybrid system.* A global cap-and-trade system can cause unexpectedly high growth in one country to reduce growth in other economies. Carbon taxes and hybrid policies are not vulnerable to this effect. However, carbon taxes tend to exacerbate economic downturns while a hybrid policy could be designed to avoid doing so.

Conclusion

The global financial crisis of 2008 has starkly emphasized a number of important lessons for the design of global and national climate policy. These lessons need to be considered explicitly during international negotiations for future climate agreement. Anticipating shocks may mean rejecting global climate policies that reduce emissions reliably in stable economic conditions but are vulnerable to collapse in volatile conditions.

Author affiliation

Warwick J. McKibbin is Professor of International Economics in the Australian National University's College of Business and Economics, and non-resident Senior Fellow at the Brookings Institution.
Adele Morris is a fellow and Deputy Director for Climate and Energy Economics at the Brookings Institution.
Peter J. Wilcoxen is Associate Professor of Economics and Public Administration at the Maxwell School of Syracuse University, and Nonresident Senior Fellow at the Brookings Institution.

Section VII

Epilogue

Richard Schmalensee

Overview

Overview

History's evaluation of this generation will depend to an important extent on its handling of the climate problem – not just on what gases are left in the atmosphere, but also on what durable climate policy architecture is left to the next generation. This paper addresses what makes the international dimension of the climate problem so difficult and important, the history of climate policy debates, and some key elements of policy architecture that those debates have so far produced.

Discussion

Climate change would be a very difficult issue even without its international dimension. The future benefits of reducing emissions are highly uncertain, and the costs of reducing emissions in the future will depend critically on the unknown pace of technological innovation. But these difficulties seem little more than academic puzzles when set against the international dimension of this problem. If the world's poor become prosperous in anything like the same way today's rich did, greenhouse gas emissions will increase substantially, and the consequences for the entire human race are likely to be extremely unpleasant. We need both to show the poor countries a much more climate-friendly path to prosperity and to induce them to follow it.

The climate policy problem has been on the world's agenda since at least the creation of the Intergovernmental Panel on Climate Change (IPCC) in 1988. In this early period some participants in the climate debate called for substantial near-term emissions reductions, but most analysts argued that it would be more efficient to focus for at least a decade on studying the climate system and developing new technologies that could reduce the costs of emissions reductions and of adaptation to climate change. This preparatory investment would permit subsequent policies both to better reflect actual risks and benefits and to impose lower net social costs.

Since this early period there has been considerable movement toward developing an international climate policy architecture. But the fundamental, critical problem of inducing poor nations to follow climate-friendly paths to development has not yet been effectively addressed. And the early-period arguments that we could wait a decade or more before making significant cuts in global emissions have passed their sell-by dates: we have in fact waited nearly two decades, and most knowledgeable observers now contend that the time for serious action is upon us – ready or not.

Key findings and recommendations

- *Climate policy will be a global concern for centuries.* It is not likely that our generation will create an international climate policy architecture that will remain workable in all its details for even a single century. It is of course essential to focus on policy designs that can be useful today, even if they fall short of what future, more stringent mitigation efforts may require. On the other hand, the core elements of policy architectures once put in place are not easily changed.
- *Architectural elements that will be of enduring value in the Framework Convention on Climate Change (1992) include the coverage, in principle at least, of all anthropogenic greenhouse gases and of all sources and sinks thereof.* In addition, the Convention and the Kyoto Protocol (1997) properly stress the importance of measurement of sources and sinks and call for the creation of what seems on paper, at least, an appropriate institutional structure.
- *But some necessary elements are missing, and some elements that are present will need to be excised or worked around in the future.* There is little reflection of the critical need to develop *new* technologies for measurement of sources and sinks, for reducing net emissions, and for enhancing the ability to adapt.
- *Also, the Clean Development Mechanism (CDM) plainly needs to be transformed or at least fundamentally reformed.* Perhaps the CDM can be transformed into a useful mechanism for technology transfer or for some other purpose beyond limiting emissions.
- *The most serious problem with the architectural elements currently in place, however, is the "deep, then broad" approach they dictate.* The Convention

divides the world into Annex I nations with emissions reduction obligations and those without. This division is at best an imperfect reflection of relative incomes when the Convention was drafted.

• *Serious US action seems a necessary condition for substantially broadening international participation in emissions mitigation efforts, but it will not likely be sufficient.* Since the climate problem cannot be solved without the participation of poor nations, particularly India, China, and other large and growing countries, it is thus critical to explore ways to modify the current architecture in ways that encourage their participation.

Conclusion

The most important and difficult climate change task before the world's policymakers today is not to negotiate Annex I emissions limits for the immediate post-Kyoto period, nor even to design the policy regime for that period. The most important and difficult task is to move toward a policy architecture that can induce the world's poor nations to travel a much more climate-friendly path to prosperity than the one today's rich nations have traveled.

Author affiliation

Richard Schmalensee is the Howard W. Johnson Professor of Economics and Management at the Massachusetts Institute of Technology's Sloan School of Management.

Appendix 2 Selected List of Individuals Consulted
Harvard Project on International Climate Agreements

We wish to thank the following individuals–and many others–who took time from their exceptionally busy schedules over the past two years to meet with us and offer their observations and insights in regard to the work of the Harvard Project on International Climate Agreements. The Project has benefitted tremendously from these meetings and exchanges. However, none of the individuals listed have reviewed, let alone approved, the content of this volume. Institutions listed are for identification purposes only and were accurate at the time of our consultation(s).

Jan Adams
First Assistant Secretary and Ambassador for Climate Change, Department of Climate Change, Government of Australia

His Excellency Ban Ki-moon
Secretary-General of the United Nations

Kit Batten
Senior Fellow, Center for American Progress

Peter Betts
Director, International Climate Change, Department of Energy and Climate Change, Government of the United Kingdom

James Connaughton
Chairman, Council on Environmental Quality, Executive Office of the President, Government of the United States

Fulvio Conti
Chief Executive Officer and General Manager, Enel SpA

John Deutch
Institute Professor, Massachusetts Institute of Technology

Stavros Dimas
Commissioner for Environment, European Commission

Elliot Diringer
Vice President, International
Strategies, Pew Center on Global
Climate Change

Robert Dixon
Team Leader, Climate Change
and Chemicals, Global
Environmental Facility

Paula Dobriansky
Under Secretary, Democracy and
Global Affairs, Department of
State, Government of the United
States

David Doniger
Policy Director, Climate Center,
Natural Resources Defense
Council

Brian Flannery
Manager, Environment
and Strategy Development,
ExxonMobil

Christopher Flavin
President, Worldwatch Institute

Jody Freeman
Professor of Law, Harvard Law
School

Masahisa Fujita
President and Chief Research
Officer, Research Institute of
Economy, Trade, and Industry
(Japan)

Kristalina Georgieva
Vice President and Corporate
Secretary, World Bank Group

Al Gore
Former Vice President of the
United States

C. Boyden Gray
Special Envoy for European
Affairs and Eurasian Energy
Mission of the United States to
the European Union

Han Wenke
Director General, Energy
Research Institute, National
Development and Reform
Commission, People's Republic
of China

Gary Hart
Wirth Chair Professor, School
of Public Affairs, University of
Colorado Denver

Connie Hedegaard
Minister of Climate and Energy,
Government of Denmark

Jim Heyes
Environmental and Corporate
Governance Officer, Global
Environment Fund

Trevor Houser
Visiting Fellow, Peterson Institute
for International Economics

Steve Howard
Chief Executive Officer, The
Climate Group

Michael Jacobs
Special Advisor to the Prime
Minister, United Kingdom

Dale Jorgenson
Samuel W. Morris University
Professor, Harvard University

Lars Josefsson
President and Chief Executive
Officer, Vattenfall

Peter Kalas
Advisor to the Prime Minister,
Czech Republic

Nathaniel Keohane
Director of Economic Policy
and Analysis, Climate and Air
Program, Environmental Defense
Fund

Melinda Kimble
Senior Vice President, United
Nations Foundation

Fred Krupp
President, Environmental
Defense Fund

Brice Lalonde
Ambassador for Climate Change
Negotiation, Government of
France

Jonathan Lash
President, World Resources
Institute

Kevin Leahey
Managing Director, Climate
Policy and Economics, Duke
Energy

Li Liyan
Deputy Director, Office of
the National Coordination
Committee on Climate Change.
People's Republic of China

Bo Lidegaard
Permanent Under Secretary of
State, Office of the Prime Minister
of Denmark

Christine Loh
Chief Executive Officer, Civic
Exchange

Lu Xuedu
Deputy Head, Office of Global
Environmental Affairs, Ministry
of Science and Technology,
People's Republic of China

Ichiro Maeda
Deputy General Manager, Tokyo
Electric Power Company

Alden Meyer
Director of Strategy and
Policy, Union of Concerned
Scientists

John Morton
Managing Director, Economic Policy, Pew Charitable Trusts

Fernando Napolitano
Managing Director, Booz and Company

Marty Natalegawa
Permanent Representative to the United Nations, Republic of Indonesia

Mutsuyoshi Nishimura
Special Advisor to the Cabinet, Government of Japan

Robert Nordhaus
Member, Van Ness Feldman

Maciej Nowicki
Minister of the Environment, Government of Poland

Marvin Odum
President, Shell Oil Company

Pan Jiahua
Executive Director, Research Centre for Sustainable Development, Chinese Academy of Social Sciences

Pan Yue
Vice Minister, Ministry of Environmental Protection, People's Republic of China

Janos Pasztor
Director, Climate Change Support Team, Office of the Secretary-General, United Nations

Annie Petsonk
International Counsel, Environmental Defense Fund

Manjeev Singh Puri
Joint Secretary (UNES), Ministry of External Affairs, Government of India

Nigel Purvis
Visiting Scholar, Resources for the Future

His Excellency Anders Fogh Rasmussen
Prime Minister of Denmark

Janusz Reiter
Ambassador for Climate Change, Government of Poland

Theodore Roosevelt IV
Managing Director, Barclays Capital

François Roussely
Chairman, Credit Suisse, France

Masayuki Sasanouchi
Senior General Manager, Carbon Management Group, Toyota Corporation

Phil Sharp
President, Resources for the
Future

Kunihiko Shimada
Principal International
Negotiator, Ministry of the
Environment, Government of
Japan

Domenico Siniscalco
Vice Chairman and Managing
Director, Morgan Stanley
International

Nicholas Stern
IG Patel Professor of Economics
and Government, London School
of Economics

Todd Stern
Senior Fellow, Center for
American Progress

Björn Stigson
President, World Business
Council for Sustainable
Development

Lawrence Summers
Charles W. Eliot University
Professor, Harvard University

Nobuo Tanaka
Executive Director, International
Energy Agency

Masakazu Toyoda
Vice-Minister for International
Affairs, Ministry of Economy,
Trade, and Industry, Government
of Japan

Koji Tsuruoka
Director-General for Global
Issues, Ministry of Foreign
Affairs, Government of Japan

Timothy Wirth
President, United Nations
Foundation

Zou Ji
Professor and Head,
Department of Environmental
Economics and Management,
Renmin University of China

Appendix 3 Workshops and Conferences[*]
Harvard Project on International Climate Agreements

Sponsored and Conducted by the Harvard Project:

Presentation to and discussion with policymakers and business and NGO leaders
Hosted by Resources for the Future, Washington, DC
October 4, 2007

Presentation to and discussion with business leaders
Venue: Harvard Club of New York City
October 18, 2007

Presentation to and discussion with scholars and stakeholders
Hosted by the Harvard Kennedy School
October 24, 2007

Workshop with stakeholders from business and the NGO community
Hosted by The Centre (policy research institute in Brussels, Belgium)
November 16, 2007

Official side-event presentation
Thirteenth Conference of the Parties, Bali, Indonesia
December 10, 2007

Workshop with Japanese industry representatives
Hosted by the 21st Century Public Policy Institute (affiliated with Keidanren), Tokyo
March 25, 2008

---

[*] This list does not include the numerous meetings the Harvard Project on International Climate Agreements has held with national climate delegations, business and NGO leaders, and many others around the world.

Presentation to scholars and government officials
Hosted by Tsinghua University, School of Economics and Management,
Beijing
March 27, 2008

Seminar discussion with scholars and government officials
Hosted by the Research Centre for Sustainable Development, Chinese
Academy of Social Sciences, Beijing
March 28, 2008

Presentations to and discussions with business leaders
Hosted by Resources for the Future, Washington, DC
September 5, 2008

Presentations to and discussions with NGO leaders
Hosted by Resources for the Future, Washington, DC
September 5, 2008

Official side-event presentation
Fourteenth Conference of the Parties, Poznan, Poland
December 6, 2008

Presentations to and discussions with senior business and NGO leaders
Fourteenth Conference of the Parties, Poznan, Poland
December 10, 2008

Presentations to and discussions with senior business and NGO leaders
Meeting of Ad Hoc Working Groups and Subsidiary Bodies of the
UNFCCC, Bonn, Germany
June 8, 2009

Major Harvard Project participation in events sponsored by other organizations:

Principal presentation to and participation in roundtable on "Architectures
for Agreement," as part of the 2007 World Energy Congress, Rome
November 15, 2007

Participation in and technical presentation at a workshop hosted by
the International Emissions Trading Association
Thirteenth Conference of the Parties, Bali, Indonesia
December 10, 2007

Participation in and background technical support for the Copenhagen
Climate Dialogue on Economic Incentives in a New Climate Agreement,
attended by senior business, government, and NGO leaders, and hosted by
Prime Minister Anders Fogh Rasmussen, Denmark
May 7–8, 2008

Participation in and technical presentations to Productivity Commission
Roundtable Discussion: Promoting Better Environmental Outcomes
Hosted by Productivity Commission, Canberra, Australia
August 19–21, 2008

Participation in and background technical support for the Copenhagen
Climate Dialogue on Role of Technology Policies in an International
Climate Agreement, attended by senior business, government, and NGO
leaders, and hosted by Prime Minister Anders Fogh Rasmussen, Denmark
September 2–3, 2008

Provided background technical presentations for a debate between energy
and environment representatives of the US presidential campaigns
Hosted by the Progressive Policy Institute, Washington, DC
September 16, 2008

Participation in and technical presentations to group of EU officials and
business and NGO leaders
Hosted by Bruegel (policy research institute in Brussels, Belgium)
September 24, 2008

Participation in and presentation to the Global Governance Conference
Hosted by Columbia University
December 1, 2008

Participation in and technical presentation at a workshop hosted by
the International Emissions Trading Association
Fourteenth Conference of the Parties, Poznan, Poland
December 9, 2008

Participation in and presentation to new Members of Congress
Hosted by the US Congressional Research Service
January 12, 2009

Participation in and presentation to the Multi-stakeholder Conference on
the Global Issues of Energy, Environment and Development
Hosted by EnergyPact Foundation, Geneva, Switzerland
March 15, 2009

Participation in and presentation to the conference on Climate Change
Policy: Insights from the US and Europe
Hosted by French-American Foundation and France-Stanford Center for
Interdisciplinary Studies
March 23–24, 2009

Appendix 4 Glossary and Abbreviations

For additional definitions, see:
 www.pewclimate.org/global-warming-basics/full_glossary
 http://unfccc.int/essential_background/glossary/items/3666.php

The texts of the United Nations Framework Convention on Climate Change and the Kyoto Protocol may be found at:
 http://unfccc.int/essential_background/convention/background/
 items/2853.php
 http://unfccc.int/kyoto_protocol/items/2830.php

Allowance	A tradable right to emit a specified amount of a substance. In greenhouse gas markets, usually denominated in metric tons of CO_2 equivalent per year. "Allowance" is generally used interchangeably with "permit".
Annex I	An Annex to the UNFCCC listing countries that would, among other things, "adopt national policies and take corresponding measures on the mitigation of climate change" (article 4, par. 2.a), upon ratification. These countries belong to the OECD or are economies in transition. The list of Annex I countries overlaps almost completely with the list of countries in the Kyoto Protocol's Annex B.
Annex B	An annex to the Kyoto Protocol listing countries that would assume legally binding commitments, upon ratification of the Protocol. Annex B also lists the actual commitments, as a percentage change in annual emissions (from which an "assigned amount" may be computed), generally from emissions in 1990. The list of Annex B countries overlaps almost completely with the list of countries in the UNFCCC's Annex I.

Assigned Amount Unit	An Annex B country's binding emissions target under the Kyoto Protocol is referred to as its "assigned amount" (Article 3, pars. 7–8). These targets are divided into "assigned amount units" (AAUs), each equal to one metric ton of CO_2 equivalent. AAUs may serve as the currency for international emissions trading under Article 17 of the Kyoto Protocol.
Banking	Saving emission permits for future use (not in the year assigned) in anticipation that these will accrue value over time. This value might be realized either through trading or through use, if emissions are anticipated to increase or become more expensive to abate.
Basket of gases	The six greenhouse gases listed in Annex A of the Kyoto Protocol, together constituting a "basket" in which Kyoto commitments are denominated. They are: carbon dioxide (CO_2), methane (CH_4), nitrous oxide (N_2O), hydrofluorocarbons (HFCs), perfluorocarbons (PFCs), and sulphur hexafluoride (SF_6).
BAU	"Business as Usual." This refers to the projected level of future greenhouse gas emissions expected without emission mitigation policies.
Benchmark	A measurable variable used as a reference in evaluating the performance of projects or actions. Also referred to as a baseline.
BTU	The British Thermal Unit is a standard measure of the energy content of fuels. It is the amount of heat needed to raise the temperature of one pound of water one degree Fahrenheit.
Bubble	An informal term referring to a provision in the Kyoto Protocol (Article 4, par. 1) that allows a group of countries to aggregate their emissions and emissions targets, and strive to meet the latter jointly. The European Union, through its Emission Trading Scheme, has utilized this provision.
Cap	An absolute aggregate emissions limit for a country, region, or subnational territory.
Cap and Trade	A policy that sets a cap for a particular air pollutant, issues permits (or "allowances" – see above) that sum

to the cap, allocates the permits to entities subject to regulation under the system (usually business firms), and provides for those entities to buy and sell the permits. The system does not, in general, specify the means by which participating entities achieve emissions reductions.

Carbon dioxide equivalent   See "CO_2e".

Carbon sequestration   The uptake and storage of carbon. Trees and plants, for example, take in carbon dioxide, release the oxygen, and store the carbon. Sometimes used in connection with CCS.

Carbon sink   Any reservoir that takes up and stores carbon. Oceans and forests are the primary carbon sinks in the Earth's carbon cycle.

CCS   Carbon capture and storage (or "sequestration") refers to a set of technologies that remove carbon dioxide from process streams in power or manufacturing facilities and store it (generally underground) for long periods of time. Only a very small number of pilot CCS facilities have been built.

CDM   Clean Development Mechanism. Article 12 of the Kyoto Protocol establishes the CDM to assist developing countries in achieving sustainable development through project-based emissions reduction. Successful projects generate "certified emission reduction" units (CERs). (In practice, the term "credit" is often used instead of "unit".) CERs may be sold to Annex I countries, which may apply them to their emission-reduction-commitments, or into carbon markets (e.g., national or regional cap-and-trade systems), where they may be traded (in some cases subject to quantitative limitations) interchangeably with government-issued allowances. The UNFCCC, through its CDM Executive Board, evaluates projects and issues CERs. An important criterion for issuance is "additionality": the Board must determine that emissions reductions resulting from the project would not have occurred in its absence.

CER   Certified Emission Reduction. One CER unit (or
                  credit) corresponds to one metric ton of CO_2-
                  equivalent emission reduction generated through a
                  Clean Development Mechanism project. (See also
                  "CDM".)
CO_2 Carbon dioxide. CO_2 is the primary greenhouse
                  gas emitted by human activities, through fossil fuel
                  combustion, land-use change, and industrial processes
                  (cement production being one of the most significant
                  contributors).
CO_2e Carbon Dioxide equivalent. The emissions of a
                  greenhouse gas, by weight, multiplied by its global
                  warming potential (see "GWP").
Commitment Period A time period during which parties to a greenhouse-
                  gas-reduction agreement are subject to the terms of
                  that agreement. For example, the Kyoto Protocol's
                  first commitment period covers the years 2008
                  through 2012.
COP   Conference of the Parties. The "supreme body of
                  the [UNFCCC]", established by Article 7 of the
                  Convention, comprised of countries that are parties
                  to the Convention. In practice, the Conference meets
                  annually, in early December.
Credit   Like an "allowance" or "permit" (which are in practice
                  nearly synonymous with each other), a tradable right
                  to emit a specified amount of a substance. "Credit",
                  though, generally refers to a right generated from an
                  emission-reduction project (usually a CER from a
                  CDM project) and may be specified more completely
                  as an "emission reduction credit" or "offset credit".
Economies in Transition The industrialized countries listed in Annex I or
                  Annex B that are undergoing the process of transition
                  to a market economy. These include some former
                  Soviet republics, including Russia, and several central
                  and eastern European countries.
EU ETS   The European Union Emission Trading Scheme.
                  The EU ETS's first trading period – which was a trial
                  period in some respects – ran from 2005 through
                  2007. The second period is coincident with the first

	Kyoto Protocol commitment period, 2008 – 2012. In late 2008, the EU specified, in part, the parameters of the third commitment period, which will start in 2013. The EU ETS is the largest and most developed cap-and-trade system in the world.
GATT	General Agreement on Tariffs and Trade. The GATT was for many years the multilateral agreement for international trade policy, succeeded in 1995 by the World Trade Organization.
GEF	Global Environment Facility. The GEF is a multilateral organization established in 1991 that provides grants to developing countries for projects that address a variety of environmental problems. The GEF is also the formally designated financial mechanism for several multilateral agreements, including the UNFCCC.
GHG	Greenhouse gas. An atmospheric gas that differentially allows incoming solar radiation to pass unimpeded and absorbs outgoing long-wavelength (infrared) radiation emitted or reflected from the Earth's surface. GHGs thereby cause warming of the Earth's surface and near-surface atmosphere. Six GHGs are named in Annex A of the Kyoto Protocol. See "Basket of gases".
GWP	Global Warming Potential. A GWP measures the effectiveness of a greenhouse gas in absorbing outgoing infrared radiation, relative to that of CO_2. The Kyoto Protocol (Article 5, par. 3) mandates the use of GWPs determined by the IPCC for comparing and aggregating greenhouse gas emissions with regard to Annex B commitments. See also "radiative forcing".
G8	Group of Eight. The G8 is a forum of the largest industrialized economies. Members are Canada, France, Germany, Italy, Japan, the United Kingdom, the United States, and Russia, which joined in 1998, after several years of informal participation. While the G8 was founded to address economic issues, it has increasingly focused on climate change policy and other matters over the last several years. Also of late, the annual summit has been preceded by a meeting

of members' environmental ministers, who have prepared the climate change agenda for the summit.

G8+5 Climate Change Dialogue   A discussion forum launched at the 2005 G8 meeting in Gleneagles, Scotland, incorporating the G8 countries and the large emerging economies: Brazil, China, India, Mexico and South Africa.

G20   Group of Twenty. The G20 includes the G8 members and major emerging market countries, providing a forum for finance ministers and central bank governors to address international finance issues.

Hot air   "Hot air" is an informal term referring to Kyoto or other targets in excess of expected (or actual) emissions, for a particular country or region. Within a cap-and-trade system, "hot air" yields excess permits that might be sold into the market. Russia and Eastern Europe have a great deal of hot air, in large part because their economies (and hence emissions) declined significantly subsequent to the 1990 baseline set by the Kyoto Protocol.

IEA   International Energy Agency. An intergovernmental organization founded by the Organisation for Economic Co-operation and Development (OECD) in 1974 that conducts analysis of energy policy and provides guidance to its member governments.

IET   International Emissions Trading, established by Article 17 of the Kyoto Protocol. Countries with Annex B commitments can participate in IET. See also "cap and trade".

IPCC   Intergovernmental Panel on Climate Change. The IPCC was created in 1988 by the United Nations Environment Programme and the World Health Organization to advise the international policy community on scientific research on global climate change.

JI   Joint Implementation. JI refers to emission mitigation projects conducted collaboratively between industrialized countries, as specified in Article 6 of the Kyoto Protocol. Such projects yield emission reduction units (ERUs) that may be traded.

Kyoto Mechanisms	The market mechanisms of the Kyoto Protocol: JI, CDM and IET.
Kyoto Protocol	A protocol to the UNFCCC that was adopted in 1997 and came into force in 2005. The protocol sets binding targets for 37 industrialized countries and the European Union for reducing GHG emissions.
Leakage	Countries that do not have binding emissions targets under an international climate agreement will gain comparative advantage in the production of carbon-intensive goods and services. A shift in production and emissions from participating to non-participating countries can result. "Leakage" refers to the shift in emissions.
Linkage	Provision for interchangeability of permits, credits, or both between and among cap-and-trade and emission-reduction-credit systems.
LULUCF	Land use, land-use change, and forestry. A potential source of emissions reductions.
L20	An analog to the G20 whose membership overlaps almost completely with the Leaders of the G20 member countries. The L20, established by former Canadian Prime Minister Paul Martin and two Canadian NGOs in 2003, addresses a variety of multinational policy issues.
MOP	Meeting of the Parties. The supreme body of the Kyoto Protocol that meets annually for negotiations, in conjunction with the UNFCCC COP.
Non-Annex I country	All countries that are not listed in Annex I of the UNFCCC, that is, developing countries and some economies in transition.
OECD	Organisation for Economic Co-operation and Development.
Offset	A reduction in greenhouse gas emissions from one source, used to compensate for (offset) emissions elsewhere. See also "credit".
OPEC	Organization of the Petroleum Exporting Countries.

PAM	Policies and measures. Under the UNFCCC, Annex I countries should undertake policies and measures to demonstrate leadership in addressing global climate change.
Permit	See "allowance".
ppm	parts per million.
ppmv	parts per million by volume.
Radiative forcing	Refers to the difference between radiation coming into the atmosphere and reaching the Earth's surface, and the radiation emitted from the Earth's surface. Positive radiative forcing increases the temperature of the lower atmosphere and the Earth's surface. Negative radiative forcing cools them. Increasing concentration of GHGs in the atmosphere increases radiative forcing. Radiative forcing is closely related to the global warming potential (GWP) of individual GHGs.
REDD	"Reducing emissions from deforestation and degradation".
RGGI	Regional Greenhouse Gas Initiative. A cap-and-trade scheme introduced in ten US Northeast and Mid-Atlantic states, beginning in 2009.
RTA	Regional trade agreement.
Targets and timetables	Targets refer to binding emission caps, and timetables refer to the timing of the commitment period during which these caps must be met.
US EIA	US Energy Information Administration, an agency of the US Department of Energy.
US EPA	US Environmental Protection Agency.
WTO	World Trade Organization.
UNFCCC	United Nations Framework Convention on Climate Change. The multilateral agreement that provides the foundation for international climate negotiations, which entered into force in 1994.